How to build a garage/barn all by yourself by building it backwards

CONTENTS:

copyright Michael D. Cahalan aka Mike Cal

this book is dedicated to Stan and Ruth
who lived in a barn house

INTRODUCTION:

A person will build something all by themselves for two reasons: 1. the extra satisfaction of an independent accomplishment, and/or 2. there isn't anyone else to help. In the mid-2000s I built a garage/barn about 95% on my own, for both those reasons.

I knew how normal house/garage construction worked: the walls are built first, then rafters are lifted up and secured in place, and the roof is sheathed. And to set the rafters in place usually take a minimum of 3-4 people. But I wanted to do it all on my own. I also wanted to set the rafters as I created them, not set all 17 at once. Yet, setting the rafters 8-9 ft in the air all by myself did not seem safe, so I decided to build the garage backwards.

Backwards meaning: I would build the roof section on the ground first, build the first level (garage walls) second, and then have a crane put one atop the other. You might think a

crane was is an unnecessary added expense, but it was only $200 / hour and the lift took less than an hour. In the end it was much easier (and safer) than having 2-3 friends and family on ladders working all day to set rafters and sheath the roof.

Over the whole build, there were just a

few instances when I did ask others for help. Maybe I could have done them myself, but having another person's help was necessary for speed and increased accuracy. And these few jobs only took 1-2 hours of their time. They were the skid loader, pouring cement, setting the roof atop the walls, and roofing.

In this book there is no overall material list. Materials are listed with each task. I will also not be listing quantities of misc nails, screws, and lumber for small tasks. I hauled everything from the lumber yard to my house in my small truck, so I purchased material piecemeal, not all at once.

The garage/barn I built was 24' wide and 22' deep (actually 23'6" wide and 21'4" deep). I call it a garage/barn because of the gambrel roof and it's red paint red.

This book is not just plans. It is the narrative of my build. Mostly because explaining my thought process, and revealing my mistakes, will help you in your build. Keep in mind that I was a relative amateur. I did not work in the building trades for 30 years. Always remember this, if I can do it, so can you.

PLANNING & FOUNDATION:

Garage size

My city has limitations on using up too much of your backyard, ie. constructing extra buildings on your property, so I was only allowed a 2-stall garage. I settled on 23'6" wide and 21'4" deep.

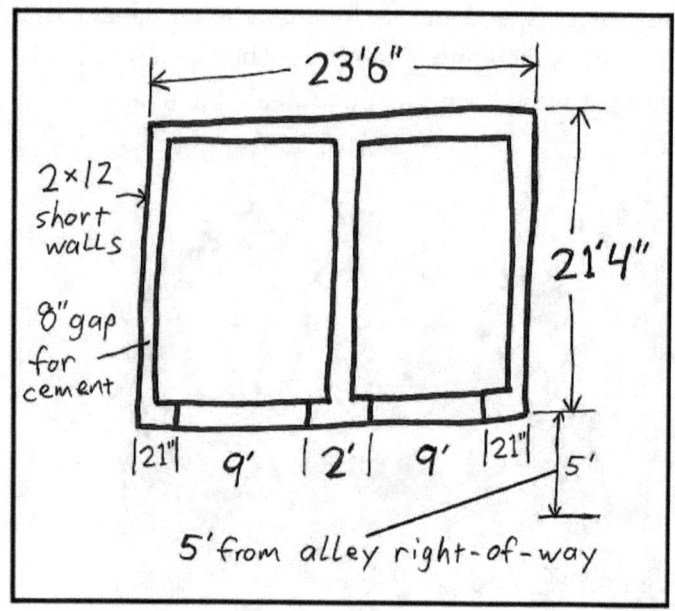

Here is why I chose these two dimensions. The rafters would run the width of the garage, which is just under 24 ft. With a support down the center, the bottom chord would be 12-ft. 12-ft lumber is pretty standard and easy to lift by oneself and not flex much when walked on. So two would make up the bottom of the rafter without much waste. If you wanted, you could stretch this out to 23'11" wide. Keeping in mind that 12ft lumber might not always be all a full 144 inches.

The depth/length was mostly dependent on the city code I mentioned before. When I got my permit, I listed 22 ft. However, 22 ft is not evenly divisible by 16" (for having rafters and studs 16" on center) so its depth is really 21'4". Initially, I thought if the depth/length was 22 ft, and the roof would have a 1 ft over hang, then the roof length would be 24 ft, I nice number for 8 ft plywood sheathing (no waste) and 3ft roof shingles.

Preparing the land

Ignoring the driveway, you should want your foundation as flat as you can get it. I was building at the back of my lot by the alley and already there were the old concrete remnants of an old one-stall garage and bushes and vines. I rented a skid loader with scoop-bucket attachment, pulled up the old concrete chunks, and flattened the area.

Tip 1: Did I haul away the old concrete? No. I put the old concrete in the lowest part of my back yard, paid to have nice black dirt delivered, spread out grass seed, and the lowest part of my back yard became the highest part of my back yard.

Tip 2: The rental place charged separately for delivering and retrieving the skid loader. I only paid for delivery, and when I was done with the machine, I drove it all the way back to the rental place late at night.

Tip 3: I didn't operate the skid loader on my property. I called a couple friends and asked, "Hey, do you wanna do something fun?" Neither had worked a skid loader before and were happy for the experience.

Layout

Exact layout or placement came next. Of course, you need to have your utilities marked. The only utility I had was a sewer line deep underground, so I had nothing to worry about.

But I made two mistakes on the placement, fortunately I caught the bigger mistake. Per city code, the garage needed to be 3 ft from the alley. I didn't want the garage to eat up any more of my backyard than it had to, so I staked out layout 3 ft from the edge of the concrete of the alley. On a whim, I double checked the plat for my neighborhood. On the plat, the alley was 20 ft wide, but when I measured the alley, it was only 16 ft wide.

So I knew 2 ft on each side of the alley was the city's land (or right-of-way), not my property. So my garage had to be 5 ft from the alley's edge. This 5 ft would be the driveway of the garage.

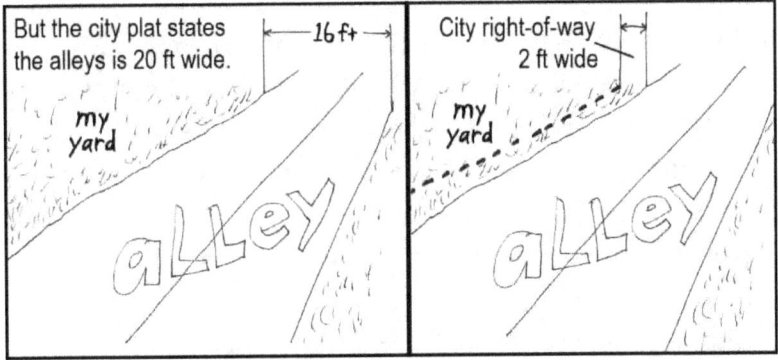

My second mistake was that, because I wanted the garage as close to the alley as possible, I kept the driveway at 5 ft. But I wish I would have pushed the garage back and made the driveway 7 ft. Because sometimes a vehicle gets parked sideways on the driveway, and 5 ft is too narrow. And 5 ft is a tight turn into, and out of, the garage. Pushing the garage back two more feet and losing 2 ft of backyard would not have ruined backyard activities.

Short walls

Many garages are built on one big slab. I wanted the garage a little higher than the ground and, when working alone, it is best to break up a large task into multiple smaller tasks. So my foundation had three pours: the short walls, the two floors, and the driveway. Also, having a short wall down the center would make it easier to float the floor since I could reach all of the floor by standing on this center wall.

I purchased twenty (20) 12ft 2x12s, 125 ft of #3 (3/8") rebar (concrete reinforcing bars), and twenty (20) 12" L bolts. The 2x12s would be the forms for the short concrete walls. The 2x12s were not wasted. After the concrete work was done, the 2x12s were used in the headers and elsewhere.

Despite the all the skid loader work, the ground was not perfectly level. I had some gravel brought in. It is alot easier to raise your building surface with gravel and rake it, than to skim off inches of dirt with a shovel.

The short concrete walls were to be 8" wide and as tall as the 2x12s. I dug a little 2-3" deep, 11" wide channel where the walls would go. So the bottom of the forms were supported by the ground/gravel itself, with just a few stakes being pounded in here and there for safe measure. Just before you pour your cement, you can spray the inside of the 2x12 forms with WD-40 (or other oil) if you wish. This will help with prying the wood forms free of the concrete later.

The tops of the forms would be supported by a 2x4 scrap, which also happened to hold the L bolts. These would be placed every 4 to 6 ft. The L bolts tie into the rebar (with wire) and are the fastening point for the eventual sill plate. The sill plate is the board (in my case a treated 2x8) that will sit atop the short walls.

2x4 scrap
holding
L bolt

rebar
hanging
from
L bolt

The rebar then hung from the L bolts, being wired together.

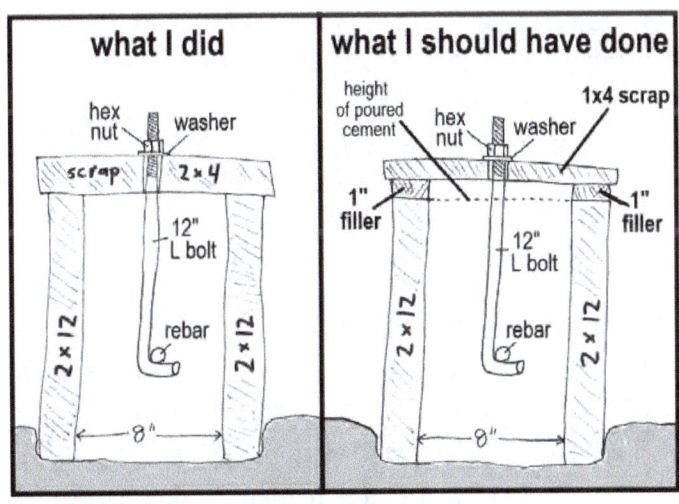

You should always have rebar or concrete wire mesh in any cement pour. A civil engineer once told me, "There are only two types of concrete: concrete that is cracked, and concrete that is going to crack." Rebar and wire will hold your concrete together after it cracks.

If I had to do it all over, I would used 1x4s to make this top support, using a filler between the support and the form.

I didn't stop the short walls at the garage door opening. I wanted the walls tied together under the eventual floor, so the short walls went from 12" to 4" at the front. These 4" walls had rebar also.

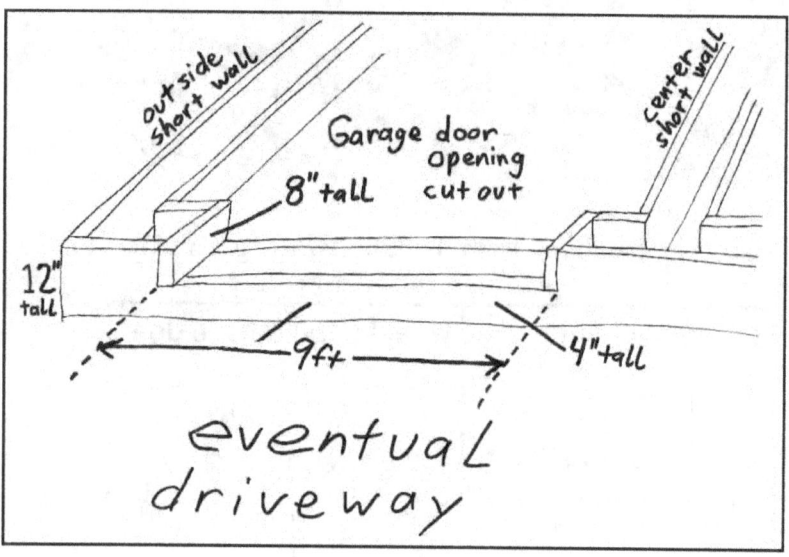

The opening for single-stall garage doors is 9 ft. A minor mistake I made was having the short walls 8" at their opening. When I installed the garage doors, the tracks would have came down onto the top of the short walls and not all the way down to the floor. So I had to use a sledge hammer and chip away at the inner outside corner of the walls to make room for the tracks. To avoid all this hard work, put some 2x4s in your forms to make some space for the eventual tracks.

concrete short wall had to be chipped away to allow garage door track to make it all the way down

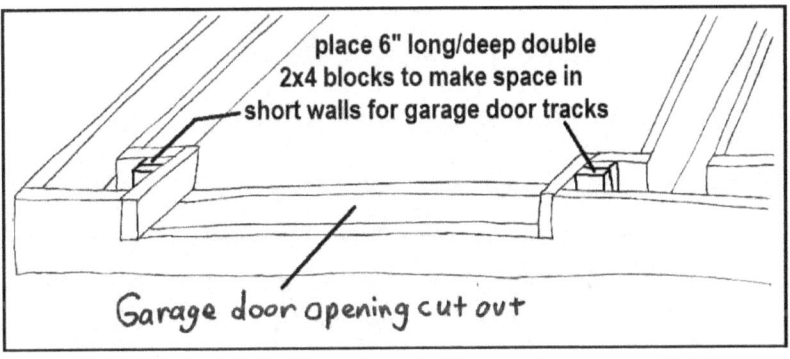

place 6" long/deep double 2x4 blocks to make space in short walls for garage door tracks

Garage door opening cut out

When you order cement from your local cement company you will need to tell them how many (cubic) yards you want. Before you can determine the cubic yards, you must calculate how many cubic feet you need.

I measured the total length of my walls that were a full 2x12 high. It came to 96 linear ft. And they were roughly 1 ft deep, but not 1 ft wide. Only 8" wide. Well, 8" is 2/3 (0.66) of 1 ft. So every linear foot of volume is not a cubic foot, it is 2/3 of a cubic foot. So I simply took 96 ft x 0.66. 96 x 0.66 = 63.4. My full-height short walls were roughly 63.4 cubic ft in volume.

The two 7 ft gaps of the garage door were 8" wide but only 4". 8" is 2/3 (0.66) and 4" is 1/3 (0.33) of a ft. So 14 ft x 0.66 x 0.33 = 3.1 cubic ft.

To turn cubic feet into cubic yards, you simply divide by 27. 66.4 cubic ft ÷ 27 = 2.5 cubic yards of cement. Adding a 1/2 cubic yard for safe measure, I ordered 3 yards of cement.

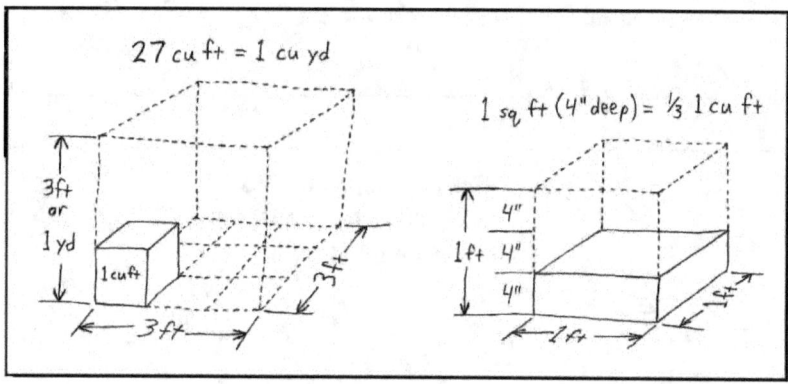

Being 5ft from the alley, it was easy for the cement truck to back in and access all the forms, and the pour went easy.

The only problem was I had to immediately pull off each 2x4 scrap piece supporting the top of the forms (and holding the top of the L bolt). So that I could then float the top of the short wall with a hand trowel where that support had been sitting. And then immediately put the support back on.

Again, if I would placed some 1x4 filler pieces under those top supports then I would have been able to float under the supports without having to pull each of them off one by one.

Floor and driveway

Pouring the floors was fairly straight forward. Days before the first pour, I had some more gravel delivered, and I spread it out. At the front, by the garage doors, the gravel was 8" below the top of the forms of the short walls (or even with the cut-out 4" wall). At the back of the garage it was only 6" down from the top of the forms. So the gravel was 2" higher in the back. That would give me a pitch of 1" for roughly 10ft, sloping down and out the garage doors. Some might suggest the gravel was unnecessary, but always remember that a cubic yard of gravel costs much less than a yard of cement. I also placed down 6" concrete wire mesh.

On the side of the concrete short walls, glue on 4" expansion joint strips or paint a guide line for cement pour of the floor

Before I poured the cement for the floors, I glued 4" wide expansion-joint strips to all the short walls. I wasn't so much worried about expansion, as these would be a guide for where I wanted the top of the floor. Instead of buying the dark fibrous strips meant for expansion joints, I simply sliced some 1/2" foam board into 4" strips. If you do not want to do this, I would suggest you paint a guide line on the inside of your short walls for the height of your floor pour.

Calculating cement order.

    3" thick (deep) floor: cubic yds = total sq ft $\div$ 4 $\div$ 27

    4" thick (deep) floor: cubic yds = total sq ft $\div$ 3 $\div$ 27

    6" thick (deep) floor: cubic yds = total sq ft $\div$ 2 $\div$ 27

    Then add a 1/2 yard for safe measure.

So the foot print of my garage was roughly 23' x 21' which is 483 sq ft. Because my floors would have concrete wire mesh, and I would not be parking anything very heavy in my garage, I went with a 4" thick floor. My cement order was $483 \div 3 \div 27 = 6$ cubic yds + extra 1/2 yard.

Again, the pour was easy because the cement truck could reach everything from the alley. Floating was also easy because we were able to walk along the short wall down the center.

A few weeks later, pouring the 5-ft driveway was the simplest of the pours. 2x4 frame on the side running from the front of the garage to the alley. 4" deep with concrete wire mesh. After this, all the concrete work was done.

CONSTRUCTING THE ROOF

Frame for the roof

Before creating any rafters, a simple frame must be built on ground near the concrete garage floor. The frame will be doubled-up 2x4s. It must be the same dimension of the outside edge of the short concrete walls, 23'6" wide and 21'4" long (or deep), with a frame down the center. This frame will also match the top of the first floor (garage walls).

The frame should be at least two 2x4 thick and needs to be as FLAT and LEVEL with SQUARE corners as humanly possible.

Construct the frame near the garage, but not too close. The mistake I made was placing my frame only 1 ft from the foundation. Later, when the garage walls were all done, I had only 1 ft of space to squeeze between the garage walls and the roof section. So place your frame at least 3 ft from the foundation.

Creating the rafters

Before I explain how I created my rafters, I want to talk about fasteners. When two boards come together, one atop the other, they commonly experience two forces trying to separate them: shear and pull-apart. I am ignoring any twisting forces.

Nails are better for shear, and screws are better for pull-apart. Nails are stronger against sheer because they are usually thicker than screws, and their lack of grooves mean less starting points for breaking failure. Screws are stronger against pull-apart because of their fastening nature (increased surface area). Simply put, you can pull out a nail with a claw hammer, but pulling out a screw is darn-near impossible. As elf carpenter Elrond would say: the screw must be unscrewed. And with nails and screws, I pre-drill holes when I can to avoid splitting wood.

And then there is wood glue. Glue does fairly equally well on shear and pull-apart forces. And glue is cheap, so I use it often. Always use exterior wood glue because you never know if the connection will get wet or not.

For these reasons, when I had two boards coming together, I would always try to use a combination of all three: screw, glue, and nail, to best combat the two forces trying to separate the boards. There is a fourth option, glued wood dowels, which I will discuss when I come to them.

Next is gussets. When two boards meet end-to-end, a gusset, or mending plate, is usually applied. I know of two types: the sharp metal gussets (truss connector plate) and DIY plywood gussets. The first are used in factories where machines press them into the wood, but you can buy them yourself and pound them in with a hammer. Not better than a factory press, but it works.

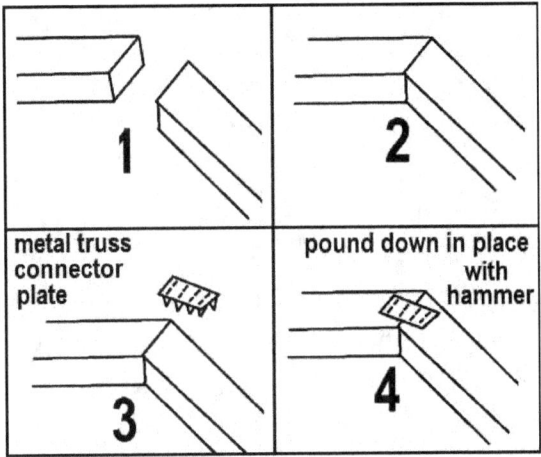

The second, you cut yourself from plywood, and glue, screw and nail into place. I used the first, to make less work for myself, but either will do.

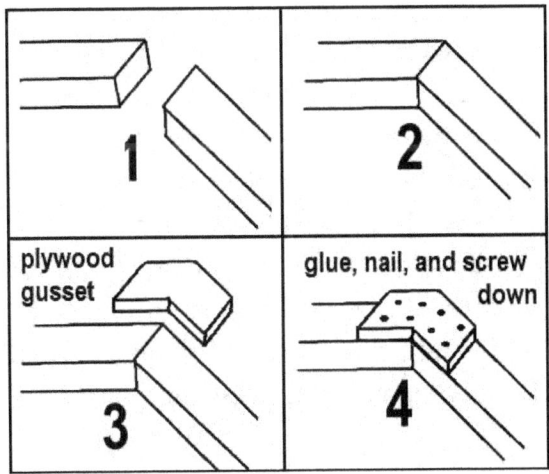

17

Another way to connect boards end-to-end is using a glued dowel. I used them on the very top of the rafter and the elbows.

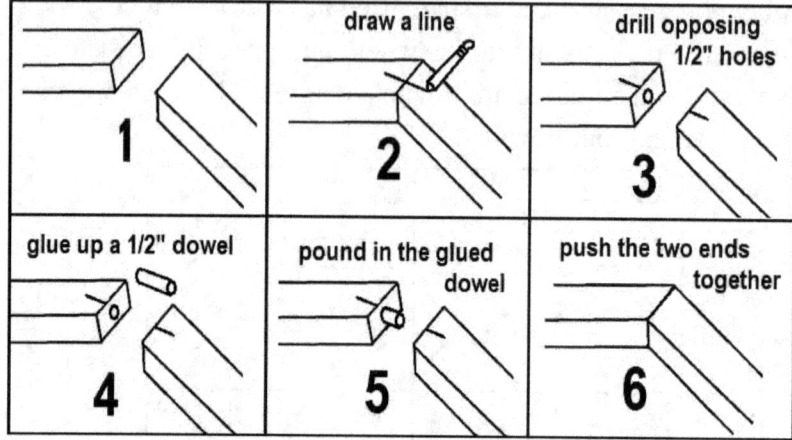

The garage foundation (concrete walls and floor) was the perfect place to create the rafters. It was flat and level, and not dirty or muddy like the rest of the yard. The material list of one rafter was: two (2) 12ft 2x8, one (1) 10ft 2x6, four (4) 8ft 2x4, two (2) 6ft 2x4, and one (1) 4ft 2x4. Some of the 12ft 2x12 used as forms for the concrete short walls can be cut up and re-purposed here.

The two 12ft 2x8s will be the bottom chords (or joists). The one 10ft 2x6 would be cut to 9'6" and be the middle king post. The four 8ft 2x4s will be the upper and lower rafters (or chords). Last, the two 6ft 2x4 will brace the elbows of the rafters, and the 4ft 2x4 will brace the top, as three collar ties.

You could use 2x12 for the bottom chords, which would be the joists for the upper level, but I knew I'd never have alot of weight up there. Just me walking around and storage for boxes and stuff. And I was concerned about the overall weight of the roof section that the crane would have to lift.

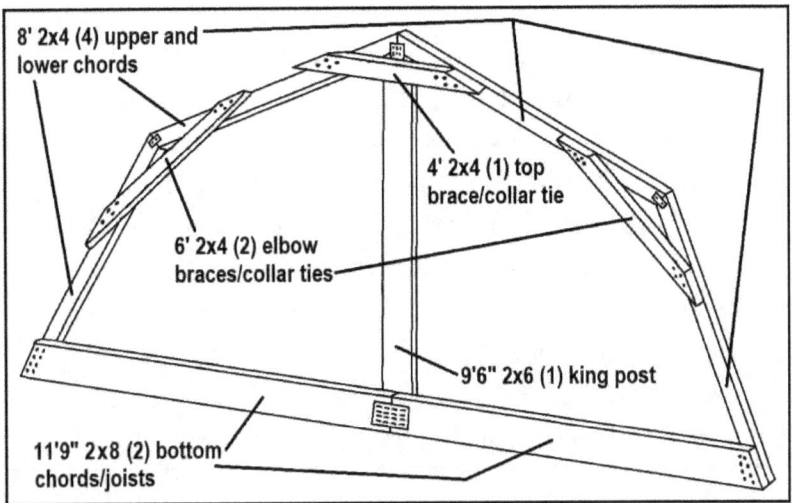

8' 2x4 (4) upper and lower chords

4' 2x4 (1) top brace/collar tie

6' 2x4 (2) elbow braces/collar ties

9'6" 2x6 (1) king post

11'9" 2x8 (2) bottom chords/joists

Here is the plan for assembling the gambrel rafters. We start with the bottom chords (joists). Place the two 2x8s end to end. They should be cut to desired length (mine were 11'9" for combined length of 23'6"). Connect them with a gusset on one side only. Then carefully flip them over. You could use dowels also, but there should not be that much force pulling them apart.

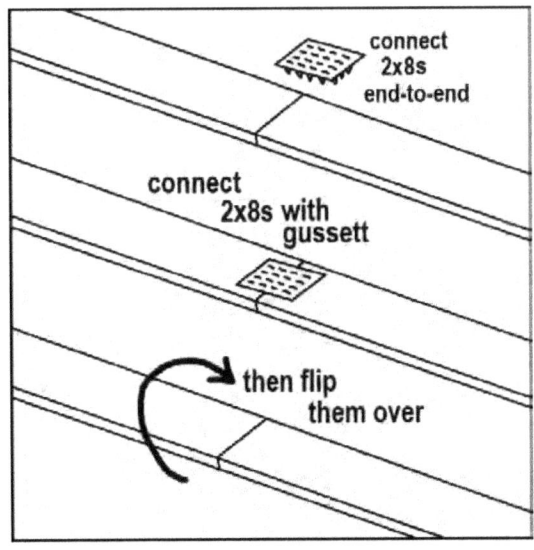

connect 2x8s end-to-end

connect 2x8s with gussett

then flip them over

19

Note, the following images show the rafter standing up. Of course, you will not be constructing it standing up. The rafter will be laying down, parallel to the ground.

Next, lay the 9'6" 2x6 middle king post atop the 2x8 bottom chords (joists) on the side without the gusset. The bottom of the 2x6 king post should be flush with the bottom of the 2x8 bottom chord (joists). The king post also must be centered and perpendicular to the joists. Measure the two diagonals from the top of the king post to the outside ends of the joists. If they are equal, then they are perpendicular.

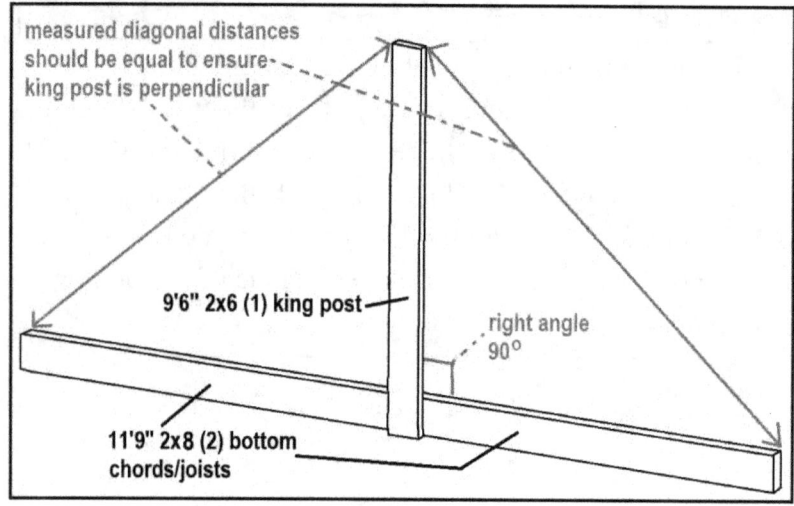

Affix the 2x6 king post to the 2x8 joists with glue and several (12-16) 3" nails and screws.

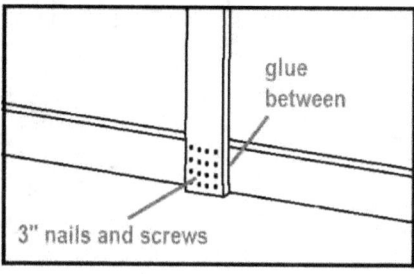

Next, a few cuts need to be made to (1) the top of the 2x6 king post, (2) the elbow ends of the four 8-ft 2x4s, and (3) the ends of the 2x8 bottom chords (joists) and bottom corner of 2x4 lower chord. I wish I could tell you the angles of the cuts to make, but the best course of action, is to lay out all the chords, draw lines, and make the cuts yourself.

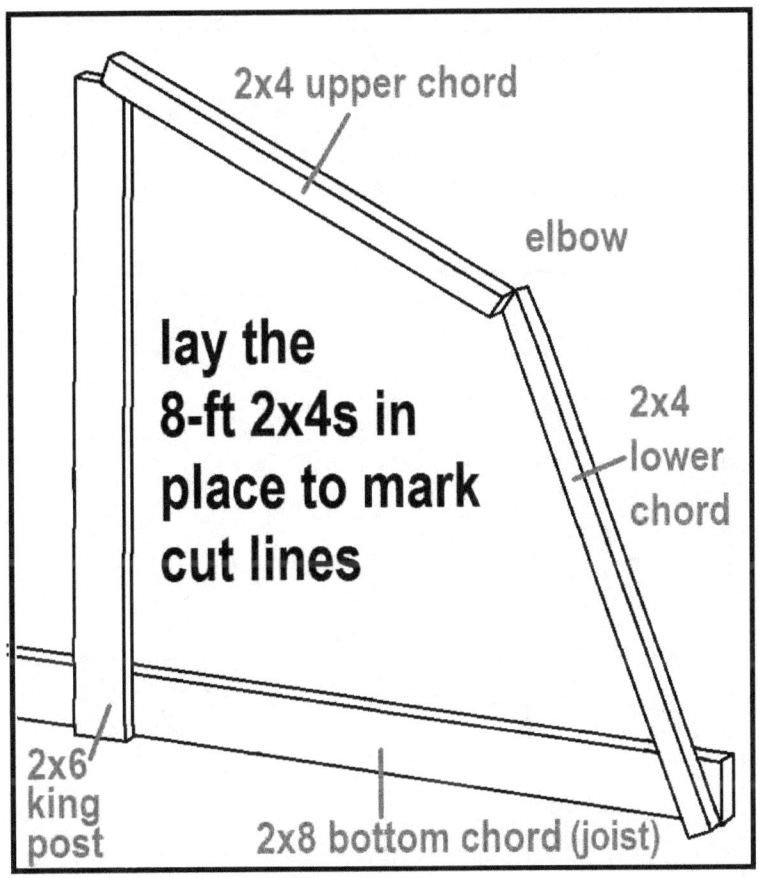

The 8-ft 2x4 upper chords will sit atop the 2x6 king post. The cuts atop the king post will be a pyramid. We lay the upper chord over the top edge of the king post to get the angle of the pyramid to cut.

First, mark the center of the king post. Second, line up the top corner of the 2x4 upper chord with the center line, while the bottom edge of the upper chord touches the peak of the king post. Draw a line under that bottom edge to get one side of the pyramid. Next, cut away the triangular piece. Then draw a symmetric angled line on the other side of the center line and cut away that side's triangular piece.

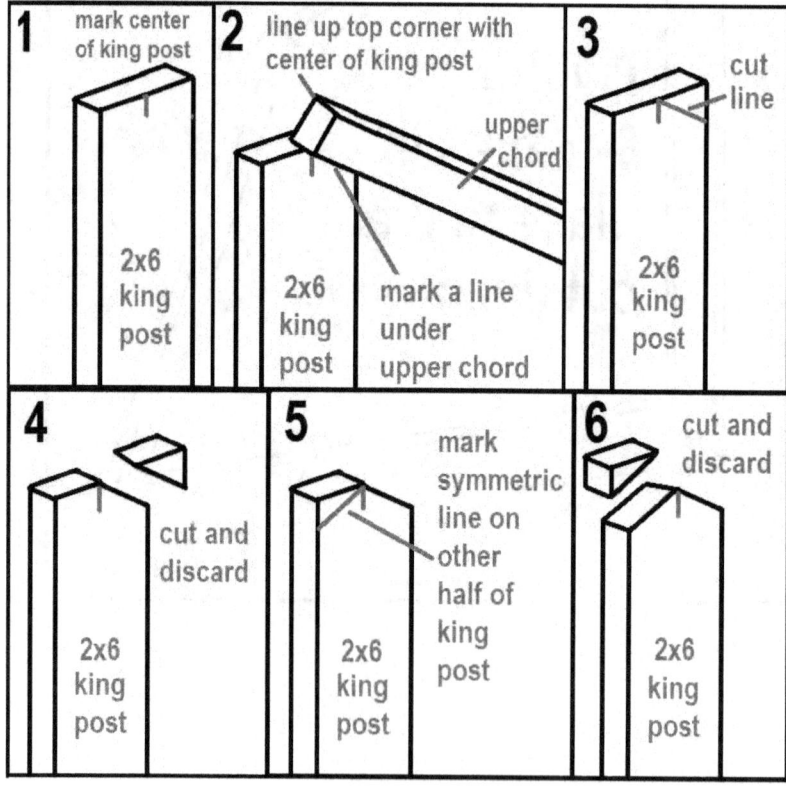

Get a scrap piece of 2x6 and make a template of the top. You will use this for making the other rafters. You cut the 2x6 king post to 9'6", then use this template to make the pyramid top.

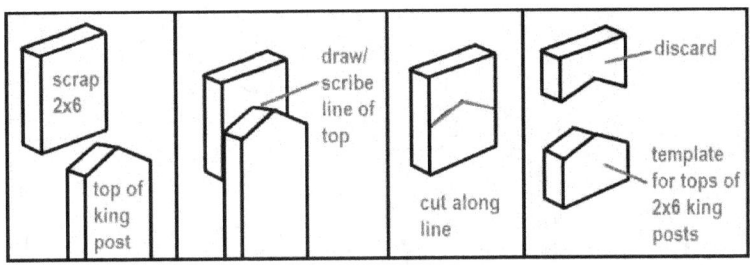

Next, again lay out the four 8-ft 2x4s in the gambrel shape. For the elbows of the upper and lower chords, lay one atop the other and cut from where they meet on the underside to the top corners of each. Your circular saw should cut clean through the top 2x4, but only partially cut (groove) the bottom 2x4. A second cut through the groove will complete the cut.

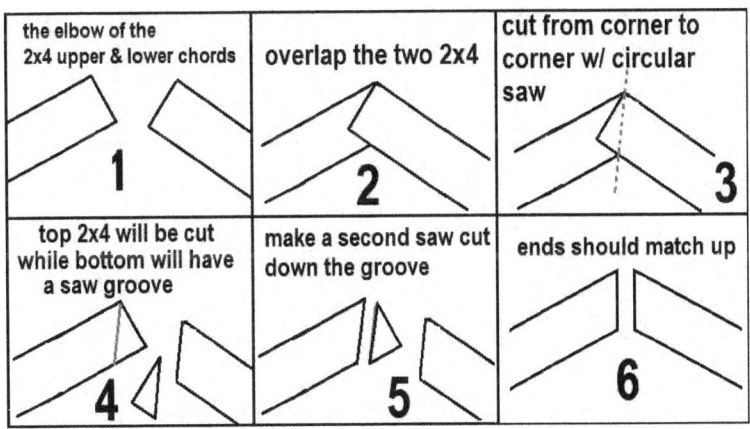

A similar type cut is made where the 2x4 upper chords come together atop the 2x6 king post.

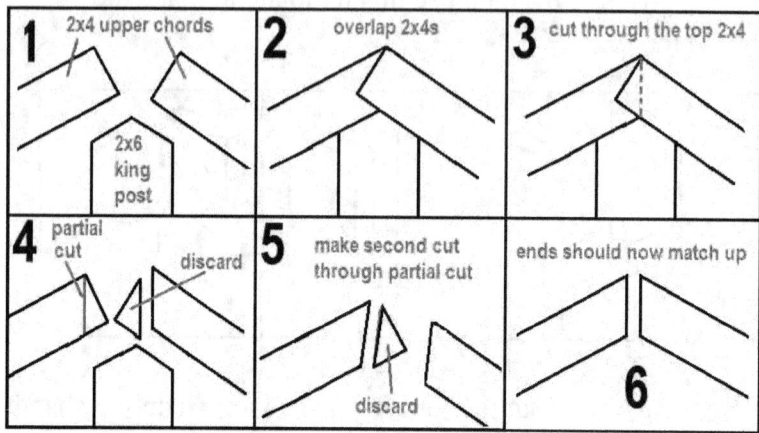

The last layout cut is where the 2x4 lower chord meets the 2x8 bottom chord (joist). Place the outer corner of the 2x4 lower chord exactly atop the lower corner of the 2x8 bottom chord (joist).

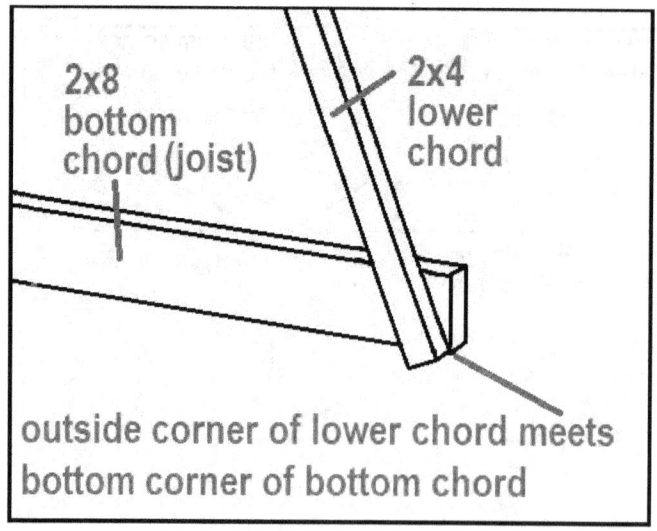

The two excess triangular portions need to be removed: the upper triangular tip of the 2x8 bottom chord (joist), and the triangular bottom of the 2x4 lower chord. You can draw lines and make complete cuts, or eye-ball them and cut them.

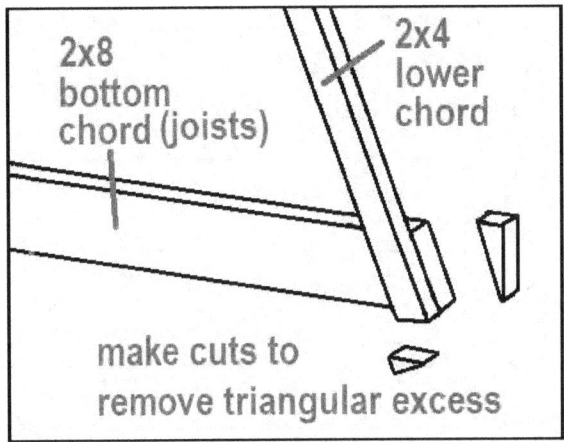

2x8 bottom chord (joists)

2x4 lower chord

make cuts to remove triangular excess

Now that all the cuts are made, it is time to make the connections for the 2x4 upper and lower chords to the cross of the 2x6 king post and 2x8 bottom chords (joists). The simplest connection was the 2x4 lower chord to the 2x8 bottom chord (joist). It just needs some glue, nails, and screws.

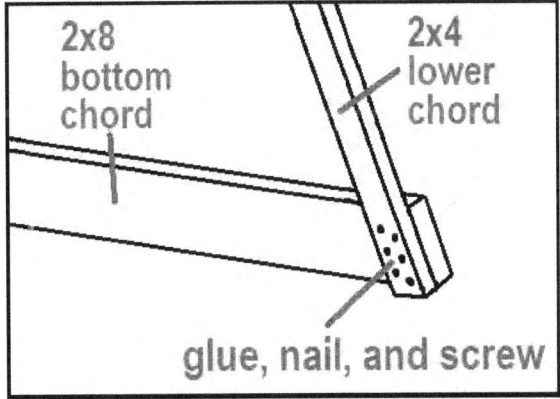

2x8 bottom chord

2x4 lower chord

glue, nail, and screw

I was a little leery about having only metal gussets to hold the elbow of the 2x4 upper and lower chords together, so I added 1/2" wood dowels. I would bring the ends together, draw a line, and drill 1/2" hole just 1" deep. Then glue up the 1/2" dowel that was just under 2" long. Insert the wood dowel and push the chord ends together. I would add a wood screw through the outside corner to hold them fast.

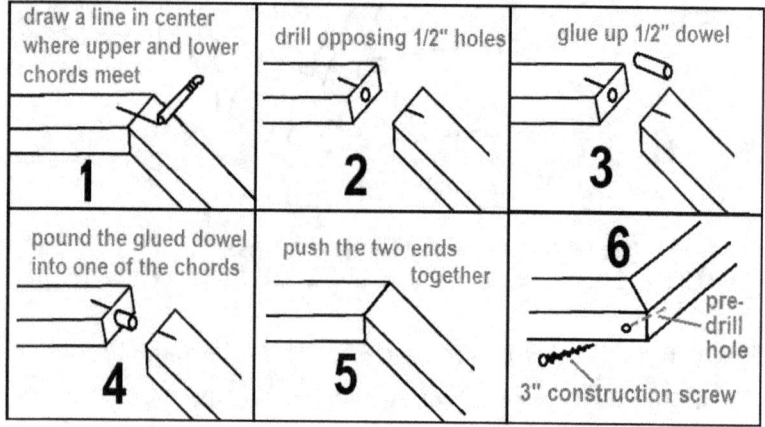

After that I pounded in the metal gussets on both sides.

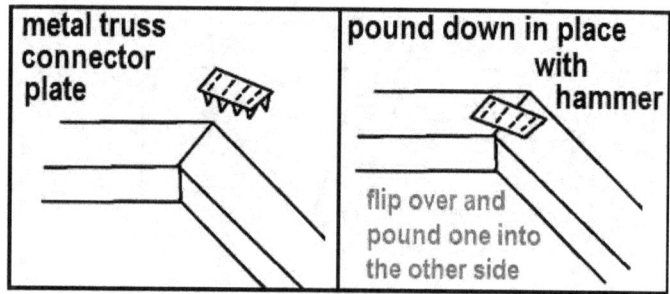

The top connection would have double 1/2" dowels. One dowel connecting the two 2x4 upper chords, and two connecting the 2x4 upper chords to the 2x6 king post. Then pound in a larger metal gusset on both sides.

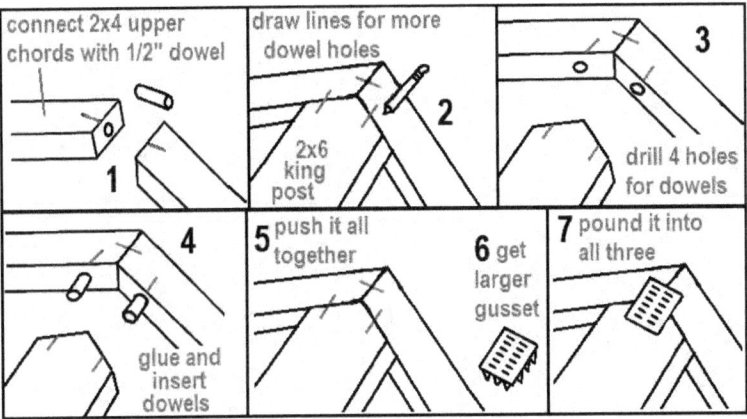

One might think the dowels are unnecessary. Maybe, for normal rafter construction, but this is not normal. These rafters must the extra strong to hold the roof together as the crane lifts the entire roof section.

Last, are the three collar ties. Notice that the 2x4 upper and lower chords, and the 2x6 king post are in one plane, and the 2x8 bottom chords are in another plane. The three collar ties will be in the same plane as the 2x8 bottom chords.

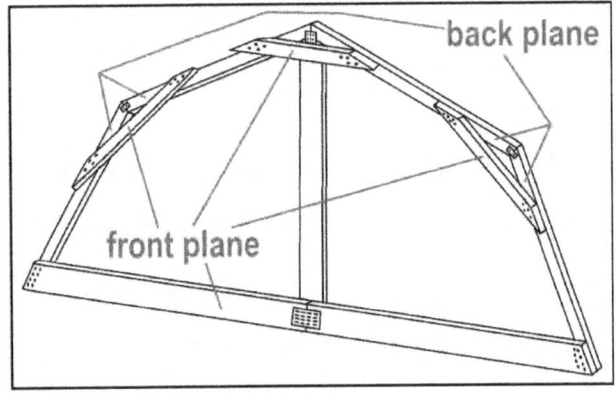

For the elbow of the upper and lower chords, simply lay the 6-ft 2x4 spanning across the chords such that as much of the 6-ft 2x4 is being used. So the bottom corners of the 6-ft 2x4 lines up with the top outside edge of the chords.

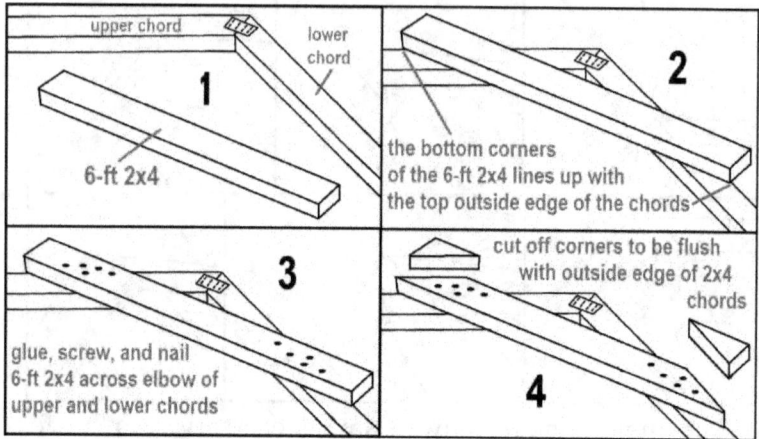

Glue, screw, and nail the 6-ft 2x4 to the 2x4 chords of the elbow. The top outside corners of the 6-ft 2x4 stick out beyond the outside edge of the rafter. So cut off these outside corners. Repeat this for the other elbow.

It is a fairly similar procedure with the collar tie connecting the tops of the two 2x4 upper chords and the top of the 2x6 king post. Draw a line down 8 inches from the very top of the rafter. Place a 4-ft long 2x4 above the line, centered with the 2x6 king post. Glue, screw, and nail the 4-ft 2x4 to the rafter. The ends of the 4-ft 2x4 stick out beyond the outer edge of the rafter. Cut off the ends so the 2x4 collar tie is flush with the outside edge of the rafter.

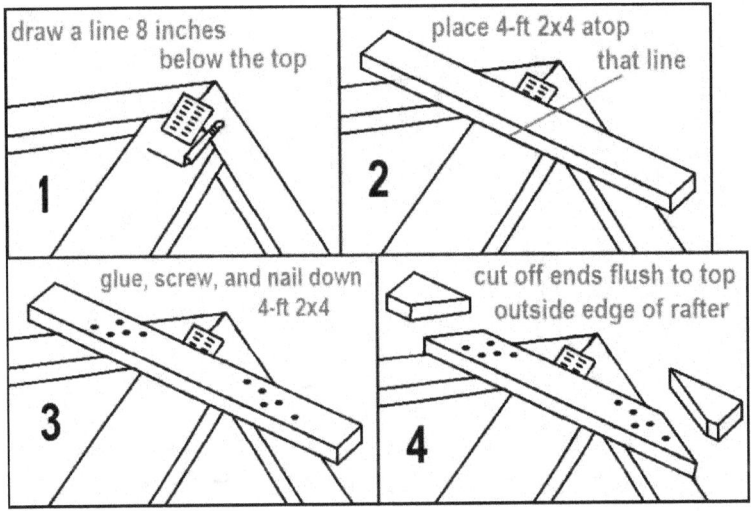

With the collar ties on, you now have a completed rafter. For my garage, I made 16 more. The first four completed rafters can be stacked / stored on the double 2x4 frame built on the ground.

Assembling roof section

Once four rafters are done, it is time to start putting together the roof section. Lift upright the first rafter and place it on the very end edge of the frame. Start on the end of the frame farthest from the garage.

When standing up, the rafters can face either direction, except for the two rafters on the two ends. Earlier I talked about the rafter having two planes: one plane with the 2x6 king post, 2x4 upper and lower chords, and another plane with the 2x8 bottom chords/joists and collar ties. For the rafters on the ends, make sure the plane with the king posts, upper and lower chords face the outside. Studs and more need to be added to these end rafters, which will be explained later.

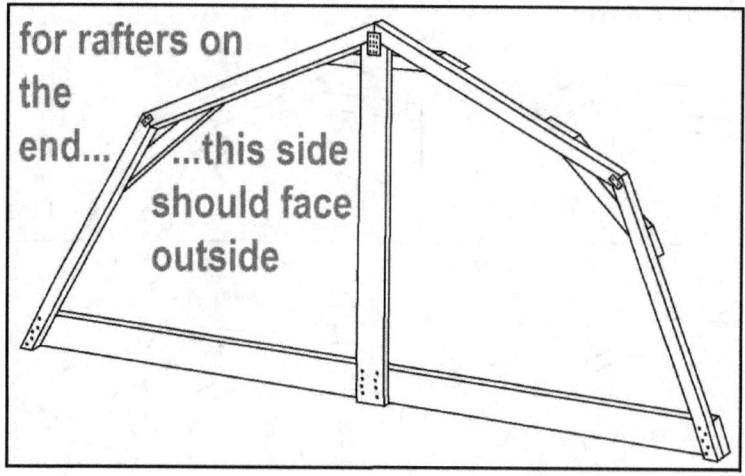

for rafters on the end... ...this side should face outside

When standing up the rafters, you will need to nail on some scrap 2x4s as crutches to hold the rafters upright. Do not nail these crutches to the outside edge of the rafter, because that is where the 4x8 plywood sheathing will be nailed on. Each rafter weighs just over 100 lbs, so this will be difficult to do by oneself, but it can be done. Make sure each stood-up rafter is vertical and plumb.

Here is a rafter next to the frame.

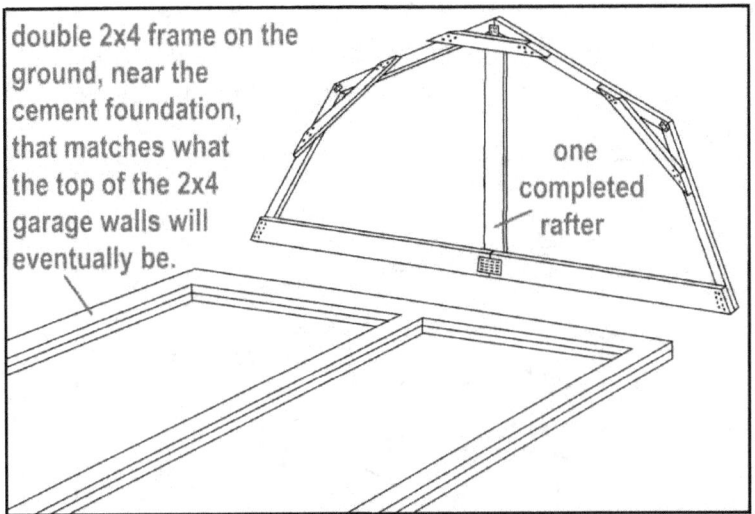

double 2x4 frame on the ground, near the cement foundation, that matches what the top of the 2x4 garage walls will eventually be.

one completed rafter

And the first rafter on the frame.

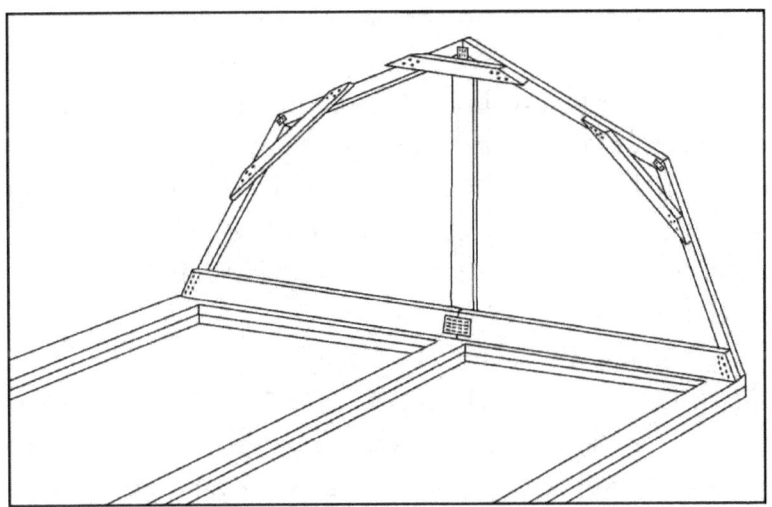

Put some 16" OC (on center) marks on the three parallel frame sections. These will be your guide when plopping down more completed rafters onto the frame. Note: the first mark will be 16" from the end of the frame, not from the center of the first rafter.

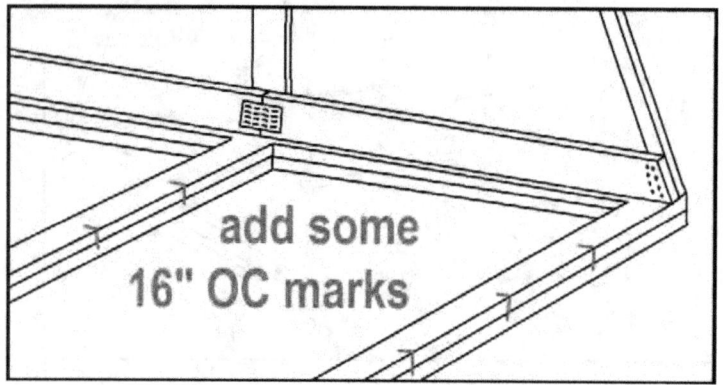

add some
16" OC marks

Let's go through the placement of all the rafters, or studs in general. When studs are placed 16" OC, an 8-ft sheet of plywood would run from the center of the first stud, to the center of the last (seventh) stud. So the outside 3/4" of the first and last studs would not be covered by the plywood. For that 3/4" to be covered, both first and last studs need to be set inward a 3/4". So at 16" OC, the gaps between the studs is normally 14-1/2", except for the first and last gaps which would be 13-3/4".

Let's add up all the distances of my garage. I had 17 rafters. The width of each rafter is 1-1/2", so 17 x 1-1/2" is 25-1/2". There are 16 gaps between the studs. 14 of those gaps are 14-1/2" wide, so 14 x 14-1/2" is 203". 2 gaps are (ends pushed in) 13-3/4", so 2 x 13-3/4" is 27-1/2".

25-1/2" + 203 + 27-1/2" = 256" = 21'4"

which is the depth/length of the garage.

Once there are four rafters standing up on the frame, you can begin nailing on 1/2" 4x8 plywood sheathing. I used plywood instead of OSB because I wanted my garage to be strong. Little did I know my garage would be hit by a flood in 2008 and a derecho in 2020. Sure, OSB lays flatter than plywood, but plywood is stronger. OSB is also less expensive, but strength is much more important, since the entire roof section must hold together during the crane lift.

Since the upper and lower chords are 8-feet long, two sheets of 4-ft wide plywood should stretch across each chord perfectly. Before sheathing, run a line or straight edge along the out edge of the rafters. If any of them are short (do not touch the line) then glue and tack on some thin wood strips to make them all level.

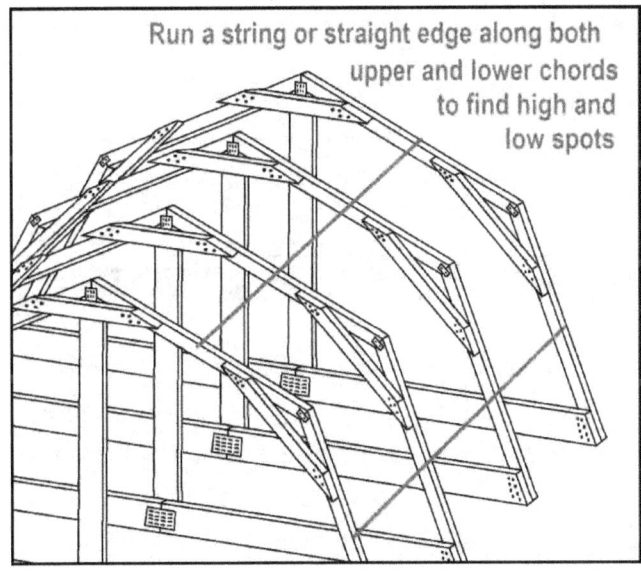

Run a string or straight edge along both upper and lower chords to find high and low spots

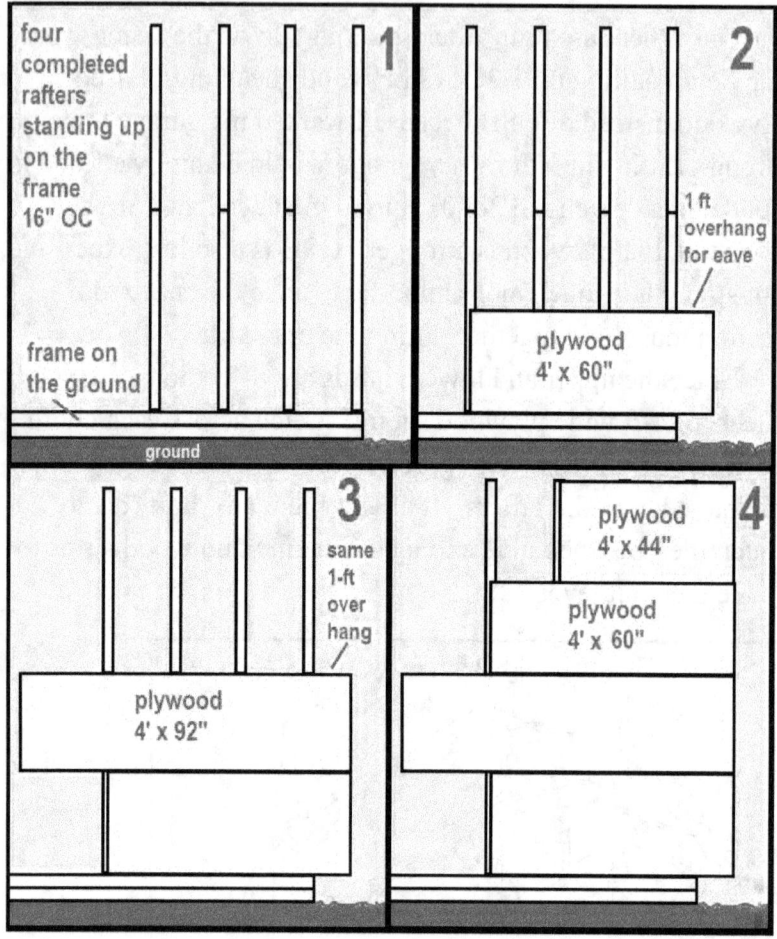

The seams of the sheathing should not line up, so alternate their lengths so there is a brick pattern. Also, make sure the sheathing has a 1-ft overhang past the first outside rafter. That is your eave. In the diagram above, the second sheet to go on is 32 inches longer than the first. So the next two rafter that will be stood up on the frame will already have a piece of sheathing over them.

Partially completed roof section sitting on the frame. Roof is covered in tarpaper.

My dad nailing down some plywood sheathing.

The first and last rafters (end rafters) are also the outside walls so they will need wall studs. Each 2x4 stud will need notches cut on both ends to get around the upper and lower chords, and bottom chords, and collar ties. Glue, screw, and nail these end-wall studs to the two end rafters.

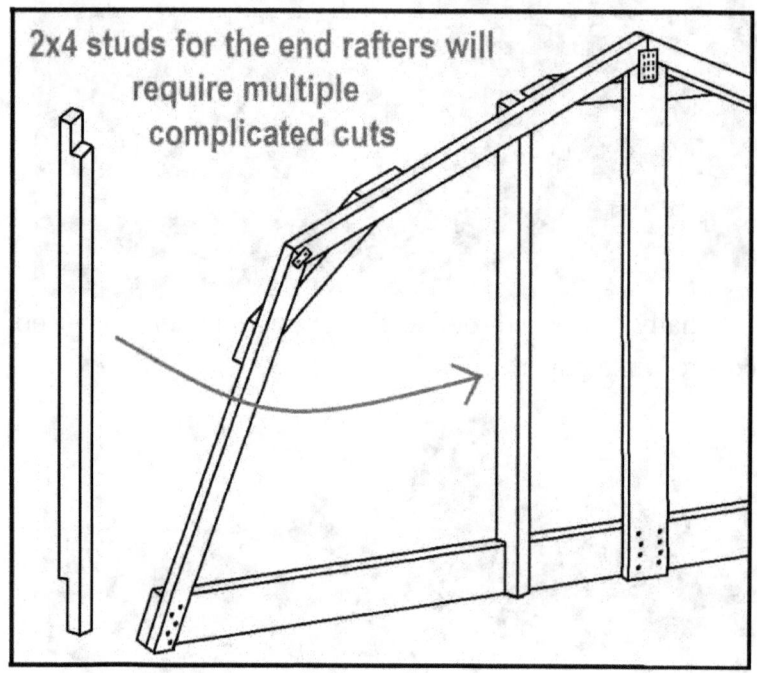

2x4 studs for the end rafters will require multiple complicated cuts

sheathing the end
start 2 ft up

You can add some outside 4x8 plywood sheathing to the end walls, but do not start at the very bottom of the wall. Leave a gap just under 2-ft. Later, when the garage wall are built, their end will leave a 2-ft gap at the top. So after the crane lift, a sheet of plywood will tie together the bottom first level (garage walls) and top roof section.

sheathing here later will
connect the top and
bottom

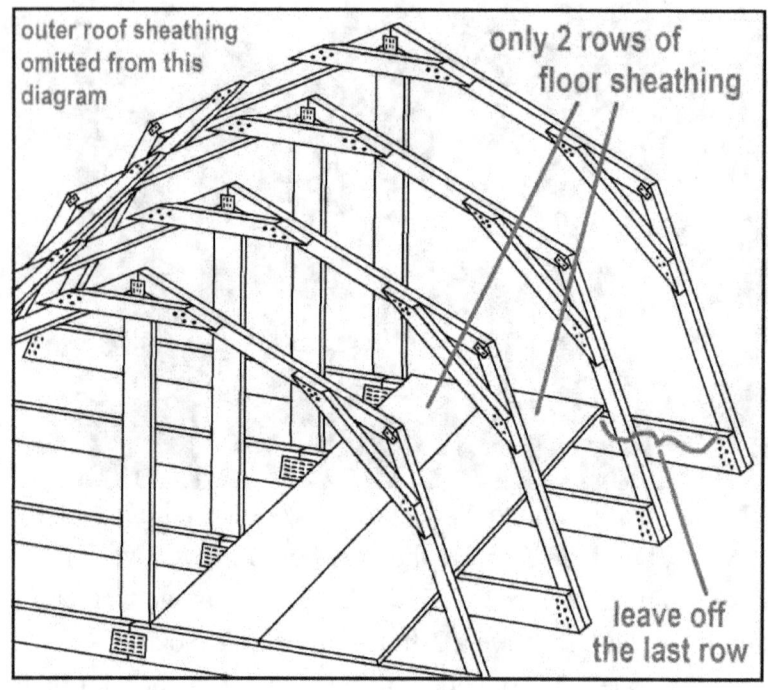

outer roof sheathing omitted from this diagram

only 2 rows of floor sheathing

leave off the last row

When, more or less, half the roof section is assembled, begin installing some flooring atop the 2x8 bottom chords (joists). Use 9/16" or 5/8" plywood. Only install two rows (8 feet) of plywood. Do not install flooring all the way from the center 2x6 king posts to where the lower and bottom chords meet. You may need that gap to see what's going on during the crane lift.

Unless you are Dick Proenneke, you are not going to get all the rafters created, put in place, and sheathed in a week, so it is going to rain on you. I covered everything with tarpaper and tarps the best I could and hoped the rain did not warp all my wood. The tarpaper I applied was temporary and tore off later.

When the entire roof section is done and sitting on the frame, it was time to start the first level (garage walls). I did not shingle the roof before the crane lift for two reasons. First, the bottom edge of the roof would eventually have flair outs which were not installed yet. Second, when I made my crane lift calculations I was confident the crane could lift the roof without the roof bending, but that was based on the weight of the roof itself and without the weight of all the shingles.

CONSTRUCTING GARAGE WALLS

Making the first level of the garage (the walls) was typical construction. Again, it is paramount that the top of the walls match the frame on the ground that holds the completed roof.

Most builders use pre-cut wall studs that are 92-5/8" so that the total wall height (with bottom and doubled top plate) is exactly 8ft. I did not care about this so all my wall studs were the full 96" (8-ft).

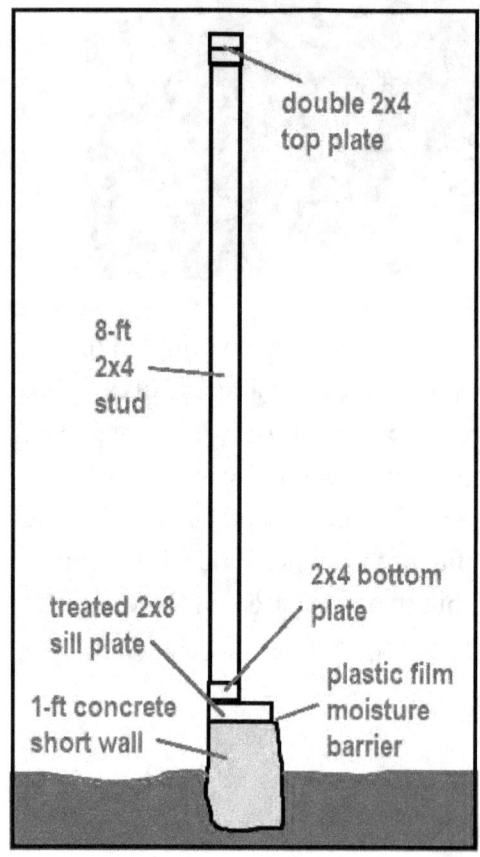

double 2x4
top plate

8-ft
2x4
stud

2x4 bottom
plate

treated 2x8
sill plate

plastic film
moisture
barrier

1-ft concrete
short wall

The linear top of the garage's concrete short walls is approximately 96 ft. Purchase enough treated 2x8 to cover the top of the short walls. 12ft lengths are ideal. These treated 2x8 are the sill plate. To hinder moisture seeping up through the treated 2x8 from the concrete walls, place 8" wide plastic film under the sill plate, or purchase a roll of the various products that are made for this purpose.

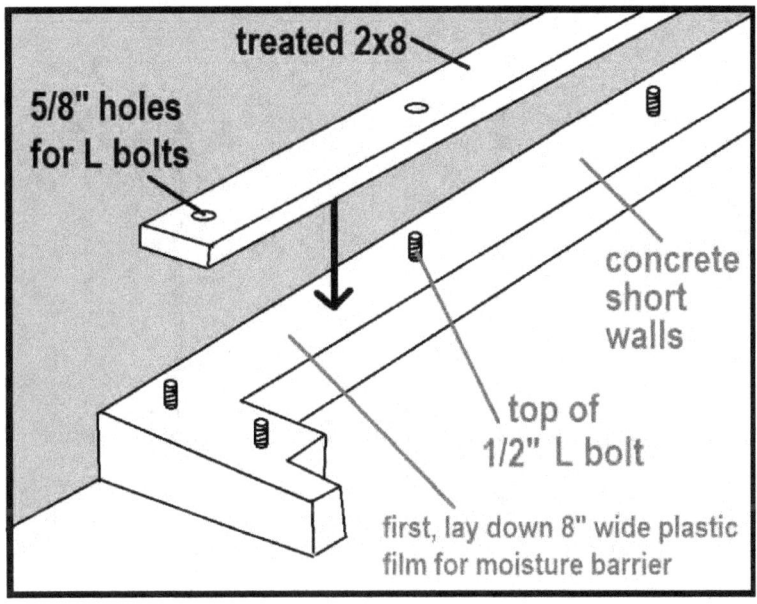

Holes need to be drilled so the sill plate can slide down over the threaded tops of the L bolts of the concrete short walls. Here is a trick: simply hold in place the treated 2x8 atop the bolts and hit the top side with a hammer. This will divot the wood and mark where the holes go. After all the holes are drilled, slide the sill plate down and add washers and nuts to the tops of the L bolts and screw down.

The stud walls are created on the ground in 8-ft sections. Then tipped up vertically into place and nailed onto the sill plate, which is atop the concrete short wall. Create another, stand it up, and butt it up to the first section atop the sill plate. Glue, screw, nail the two sections together. These stud walls only have one 2x4 along the top. So add a second 8-ft 2x4 to the top to complete the double top plate. Have that top 2x4 straddle both wall sections to help tie them together.

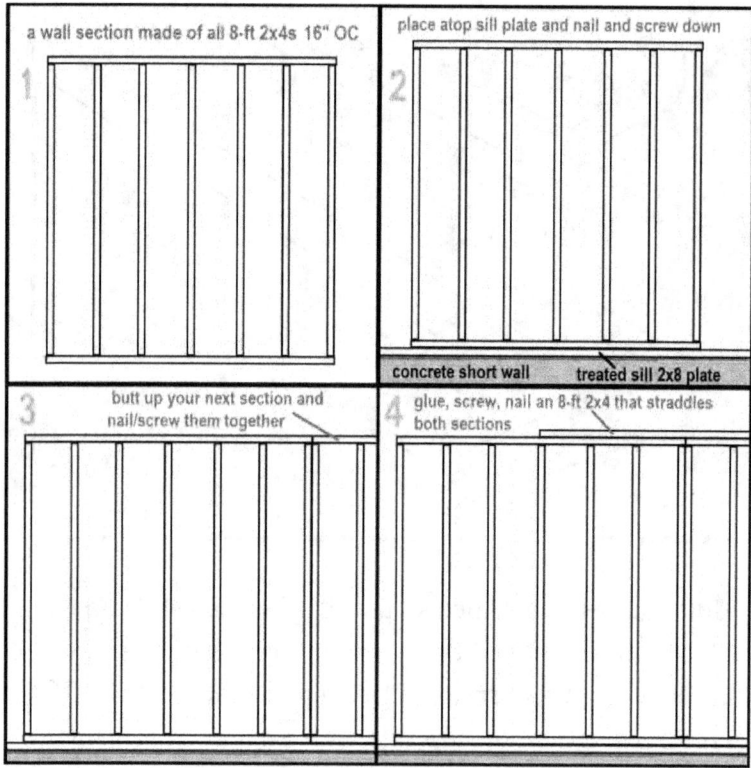

Put doors and windows where you like. I had one door in the back and two 1'x2' horizontal windows on each side. I did not cut out my windows until after the crane lift. If you want lots of windows, do not install them now. The garage walls need to have as much strength as possible, so leave the wall studs and outer sheathing complete. Cut out the studs and sheathing for windows after the crane lift.

Let's look at the opening for the garage doors.

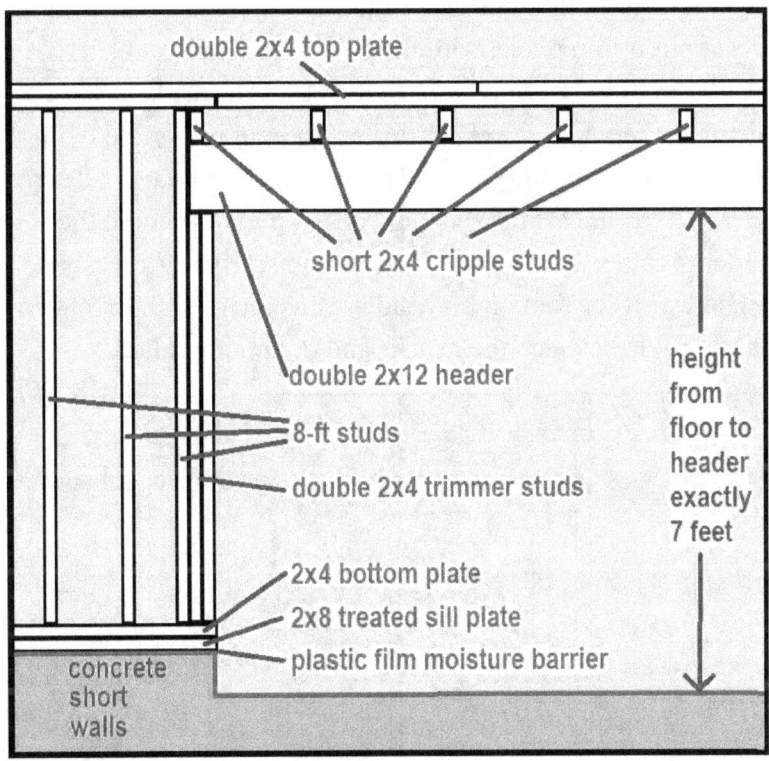

The 9-ft wide 2x12 headers above the garage door can be heavy for one person, so here is a trick. A header usually has three pieces: two 2x12s and a 1/2" thick filler piece (usually plywood or OSB), and all three are glued and nailed/screwed together. In normal construction you assemble all three together on the ground (or on saw horses) and, with the help of 1-2 more people, we lift the heavy mo-fo into place. But the header can be created by just one person without all the heavy lifting. To do it yourself, create the header up in the place it is going to sit.

First, cut all three pieces (two 2x12s, one 11" wide 1/2" plywood filler) to length. Second, lift up and put into place just one of the 2x12, atop the trimmer studs, and nail it into place. Third, slather one side of the 11" plywood filler, lift it up and put it into place, with the glue side against the 2x12 already up there, and nail it to that 2x12. Third, slather one side of the remaining 2x12, lift it up and put into place, with its glue side against the 11" filler. Fourth, then clamp all three together. Fifth, nail the second 2x12 in place and bind the header together with lots of 3" nails and screws. At least two nail/screw every foot on both sides. Last, nail in short 2x4 cripple studs between the header and double top plate.

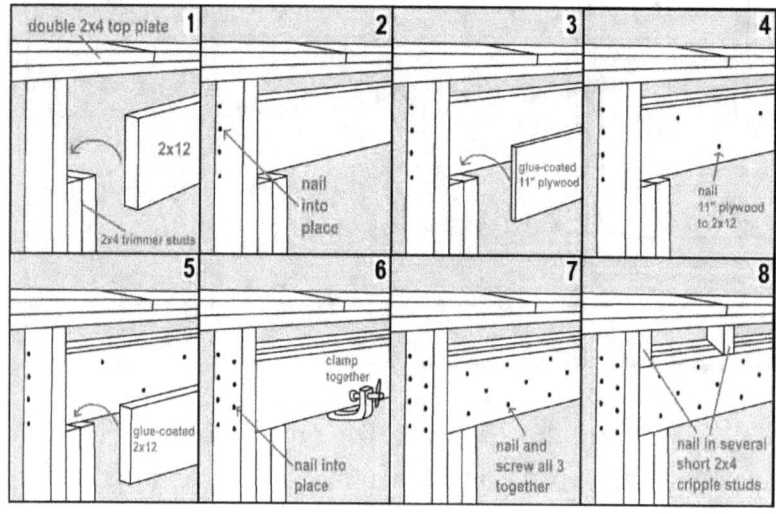

I had a stud wall down the center. If you wanted a garage that was more open, you could used headers and posts, but there must be a wall down the center. The rafters are not designed to span the entire 23'6" without support at their center. Again, the top of all your garage walls, including the center wall, must match the frame on the ground.

Next comes sheathing the garage walls. I nailed on 1/2" thick 4'x8' plywood sheathing to the outside of my stud walls. The center wall got OSB. To discourage the bottom of the garage walls from every rotting out, I started with a 2-ft first course of treated plywood around the bottom.

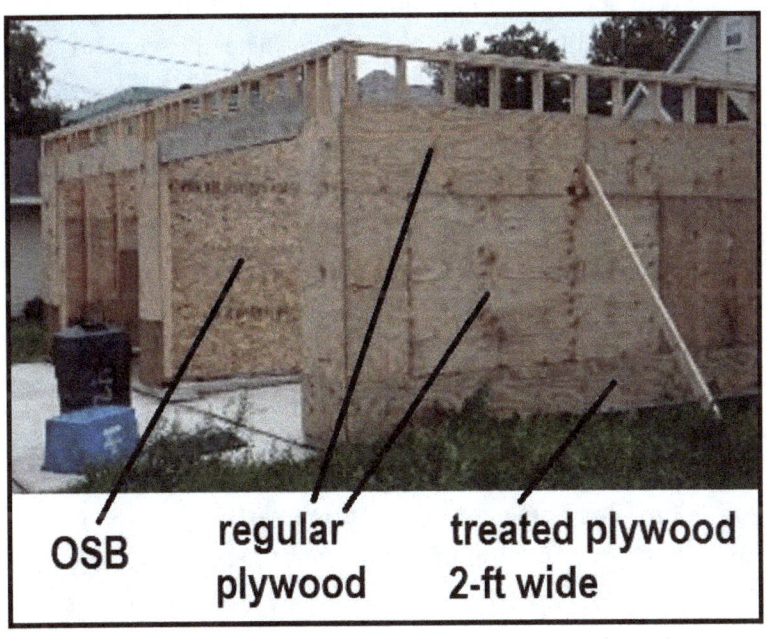

OSB | regular plywood | treated plywood 2-ft wide

Do not sheath the walls all the way to their tops on the front and back walls. After the roof is crane lifted into place, you will add sheathing to help tie the garage walls to the roof section.

After all your garage walls are done, it is time to create a backbone brace for the crane lift. This will be done up inside the roof section.

BACKBONE BRACE AND CRANE

Backbone brace

The roof section is made up of numerous rafters and sheathing. It is like a mesh framework, instead of a solid item. The crane can only attach itself to the roof section in a few spots, and any stress or movement could cause the roof section to collapse or rip apart. So there needs to be a strong stiff brace (like a handle) for the crane to attach to. It is like carrying a briefcase. You don't hold onto the briefcase. You hold on to the handle, which carries the briefcase.

So I constructed a long backbone-like brace inside the roof section. The crane will attach to this brace at four spots. In a sense, the crane is going to lift this brace, which in turn, will lift the entire roof.

Creating this brace consists of three parts: (1) adding more collar ties to the top of each rafter, (2) affixing long 2x12 runners under these collar ties, and (3) add spacing blocks to the inside of the 2x12 runners.

All the rafters, except the two end rafters will get a second top collar ties. Get 4-ft lengths of 2x4 and cut off the corners. They need to be attached at the exact same height as the collar ties already up there. If you remember, the bottom of the top collar tie was 8" from the very tip top of the rafter. So mark a line and glue, screw, nail the second top collar tie in place.

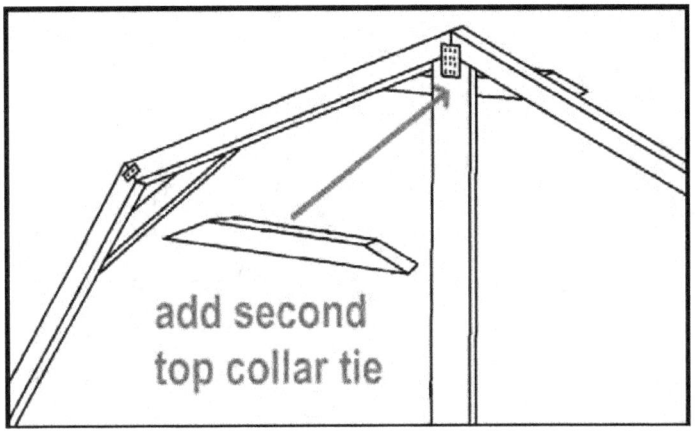

add second
top collar tie

The brace will be two sets of 2x12 (runners) stretching the length of the roof section. This will be nailed up just underneath the top collar ties, on both sides of the 2x6 king post. They are nailed into the edges of the 2x6 king post with 16p duplex (double-headed) nails. The reason duplex nails are used is so we can take down the 2x12 runners after the crane lift. You could leave the 2x12 brace up there after the crane lift, but there is no further need for them.

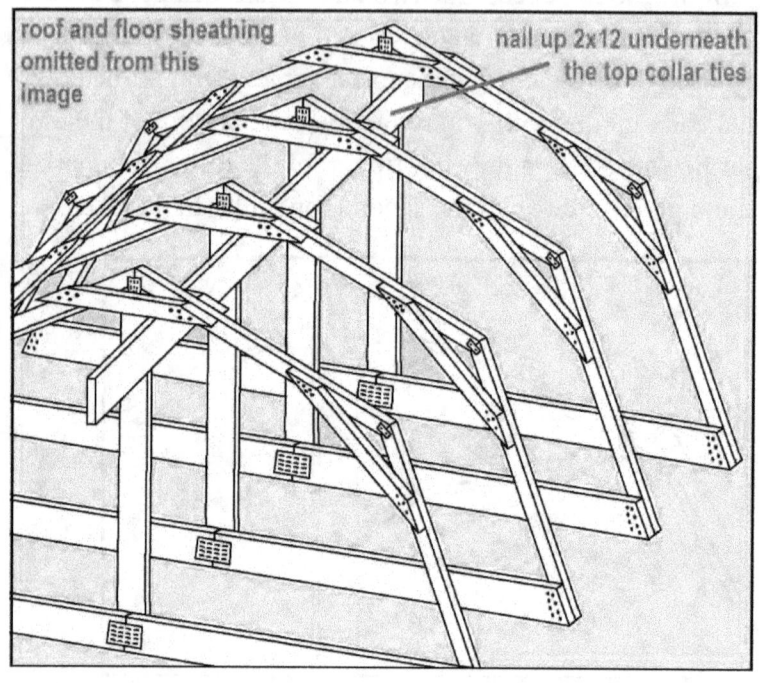

roof and floor sheathing omitted from this image

nail up 2x12 underneath the top collar ties

The image above shows just one 2x12 on one side. There will be a second one on the other side of the 2x6 king post.

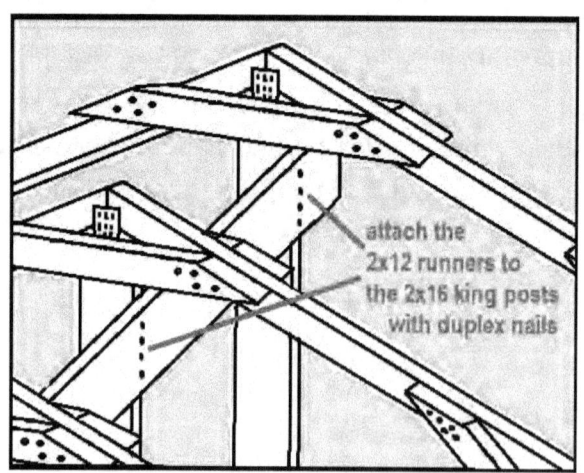

attach the 2x12 runners to the 2x16 king posts with duplex nails

The roof section is 21'4" long so four total 12-ft 2x12s will be needed. When two 2x12s come together end-to-end connect them with a 1-ft scrap 2x12 as a splice or gusset. Attach the scrap with 3" screws so it can be removed after the crane lift.

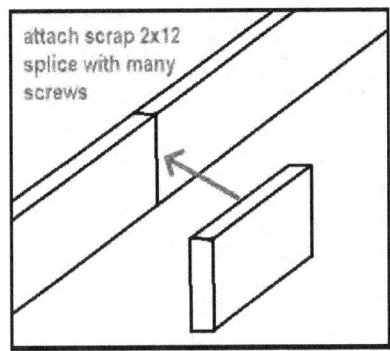

Make sure the lengths of the 2x12 runners are alternated on both sides so their ends (seams) do not match up.

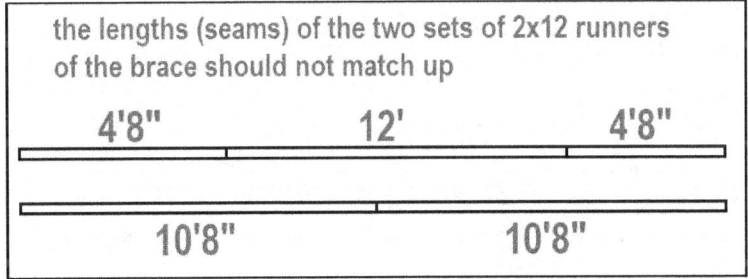

To further lock the rafters in place, several 14-1/2" 2x4 blocks are added to the inside of both 2x12 runners. These blocks wedge between and butt up against the 2x6 king posts, holding them apart during the crane lift. Again, I used screws to attach these blocks so they could be removed after the crane lift.

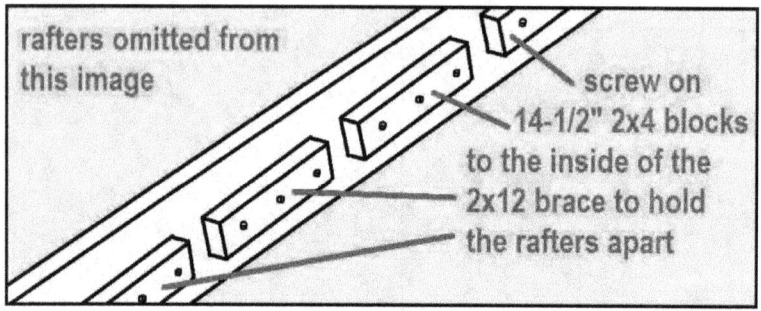

It is best to affix these 2x4 blocks after the 2x12 brace is up. This will be difficult since the two 2x12 runners are only 6" apart from each other, but it really is the best way to make sure the 2x4 blocks are tight between the rafters.

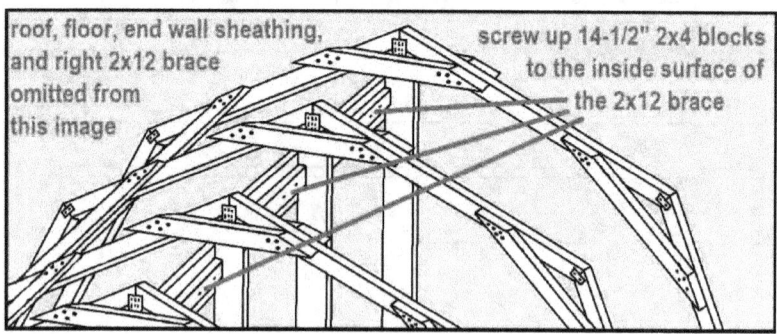

To save yourself some frustration, you could nail up the 2x12 runner for one entire side, then screw on all its blocks without the obstruction of the other 2x12 runner. Then nail up the second runner across from it, and work to get its blocks screwed in. Either way, it will be a tight working space.

Once the 2x12 backbone brace and all its 2x4 blocks are up, we are ready for the crane lift.

Before instructions on the crane lift, I want to go through my calculations for the weight of the roof section and the strength of the backbone brace.

Each rafter has approximately 52 linear feet of 2x4 (upper and lower chords and collar ties). I used 1.25 lbs per linear foot for 2x4. 10 linear feet of 2x6 (king post) at 1.5 lbs per ft. And 24 linear feet of 2x8 (bottom chords/joists) at 2.0 lbs per ft. So the weight of just one rafter is 128 lbs. Multiplied by 17 rafters is 2,176 lbs.

I estimated 24 sheets of 1/2" plywood to sheath the roof, and used 10 sheets for the ends. At 45 lbs per sheet their total is 1,530 lbs. Approximately 12 sheets of 5/8" plywood to sheath the floor, at 55 lbs per sheet, adds 660 lbs.

So the total weight of the roof section comes to 4,366 lbs. We could add in some odds-and-ends and estimate the total at 4,500 lbs. If that is divided by the 17 rafters, then the backbone brace must hold every 16" the roughly 265 lbs at each rafter. I do not know the specific strength limits of 2x12s, but I figured two of them could hold 265 lbs.

The other contact point is the 4 places the crane will attach to the brace. So this time we divide 4,500 lbs by 4 which is 1,125 lbs. Again, I assumed a crane strap would not rip through two 2x12s with that much weight.

51

Crane lift

    Call or visit your local crane company and set a lift date. I doubt your crane company will have ever attempted this before (lifting an entire roof section for a "regular joe" home owner) so invite them out to look at what you have done. Specifically, show them the backbone brace that the crane will hold onto. If they are hesitant to make the lift, tell them to buy a copy of my book. Also, ask the crane company about "wind delay" days. If the day of your lift has wind over 10 mph, I would postpone to another day. Also make sure electric lines and tree limbs are out of the way.

    When the crane arrives for the lift, the best place for it to sit is next to the first level (garage walls). The crane must have a spreader bar approximately 12-ft long with two nylon straps at each end. Cranes usually bring their own straps and they are rated above 5,000 lbs and are more than good enough for this lift.

my father on the right

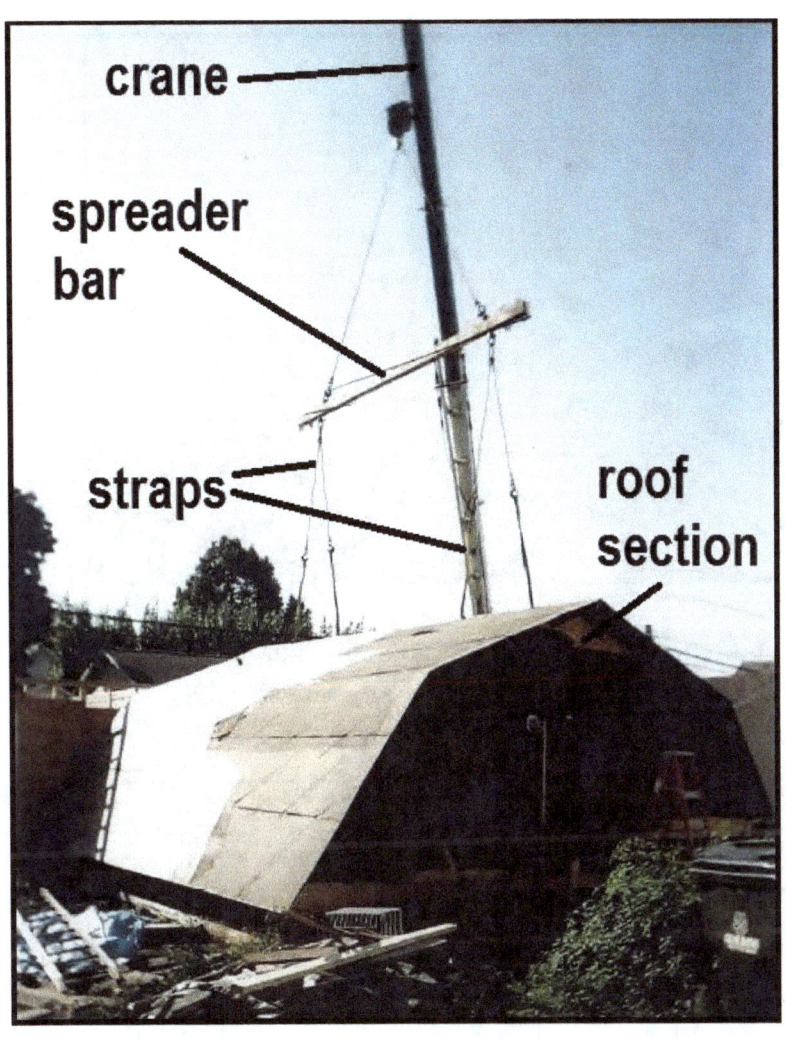

Have the crane operator lift the bar with straps over the top of the roof section. You will then climb up atop the roof section with a reciprocating saw and cut four 3" holes in the very top ridge of the roof. Where you cut your holes will depend on where the two straps hook onto the spreader bar. For my lift, the inner holes were 4-ft from the center (8-ft apart) and the outer holes were 4-ft out from them.

Lower the four straps through the four holes and wrap them around the backbone brace. You will need to crawl into the roof section to complete this task.

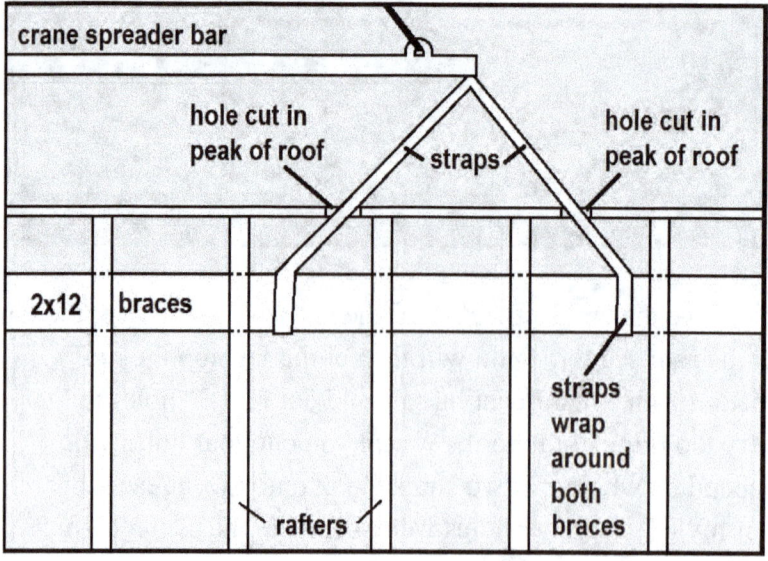

The crane can then lift the entire roof section up and over the garage walls and lower it into place.

If you can, have a few people there that day so you all can take a corner. Make sure everyone has a ladder. Don't let anyone get their fingers pinched.

The roof should fit perfectly atop your walls. Mine did not. On the back, one side was 1/2" outside the wall, and the other side was 1/2" inside the wall. So either the top of the walls, or the bottom of the roof section, was just a smidge out of square. I knew having the crane lift the roof up and setting it back down would not change anything, so I called it good.

A tall ladder is needed to get up to the top and unhook the straps, once the crane operator lowers the crane some to give some slack. And just like that, the crane lift was done. I was only charged an hour of crane time which was $225.

Before the lift discuss with the crane operator any specific words or hand signals they want from you to communicate during the lift.

Before the day ends, the roof section needs to be affixed to the top of the first level (garage walls). This is done four different ways.

1. Toe-nail the bottoms of the rafters to the walls.

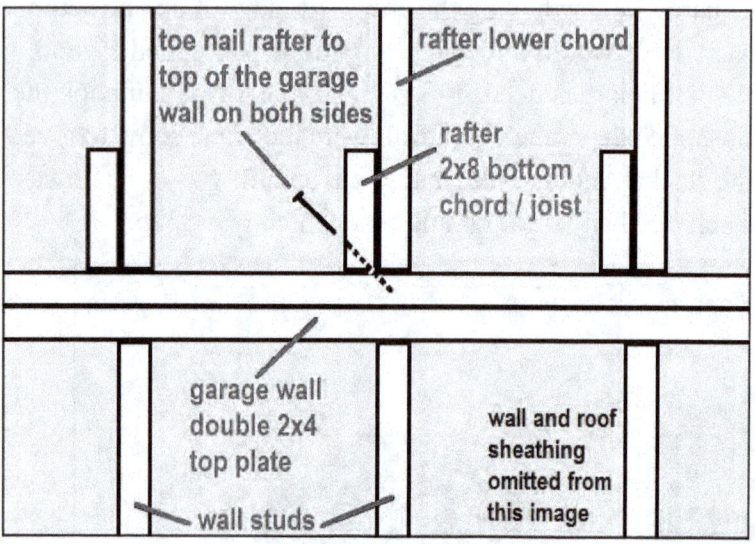

2. Nail and screw on rafters ties that connect the top plate of the garage walls to the rafter bottom chord / joist.

3. On the end walls, pound metal truss connector plates or mending plates (or plywood scabs) that connect the garage wall top plate and the rafter bottom chord (joist).

4. Finish sheathing the front and back of the garage.

finish sheathing ends with plywood to band-aid the top and bottom together

EXTERIOR EXTRAS

There are quite a few extras to add. Some of them must be completed before shingling the roof: flare outs, turkey tail, facia and blocks, and shed dormers. These all affect the roof surface so they must be done before shingling.

Flare outs

Before shingling the roof can begin, 45° flareouts need to be added to the bottom edge of the roof. The ends of the roof already have an eave, 1-ft of extra plywood. The flareout adds 1-ft to the roof's bottom edge, at an angle much flatter than the lower chord, and this overhang becomes the eave of that lower roof edge.

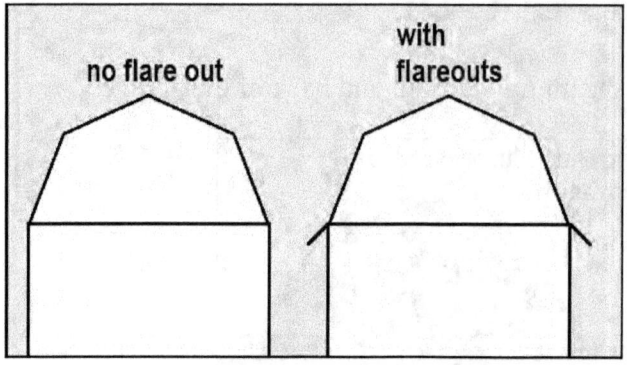

The flareouts have a 1-ft wide 1/2" plywood top with numerous triangular 2x4 supports below them. The supports I made were a simple two-piece 2x4 V that was toe-nailed to the outer garage wall.

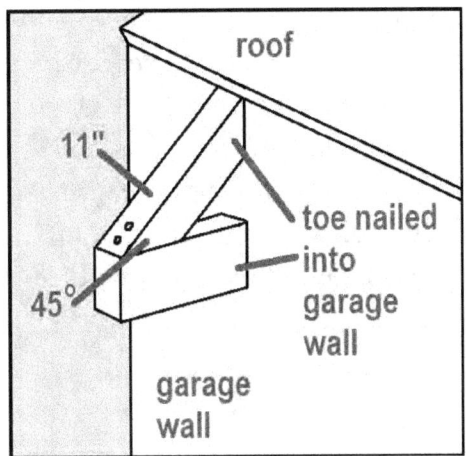

I wish I would have made a three-piece triangle with one side being a base that set against the outer garage wall. This would have made it much, much easier to attach to the garage wall.

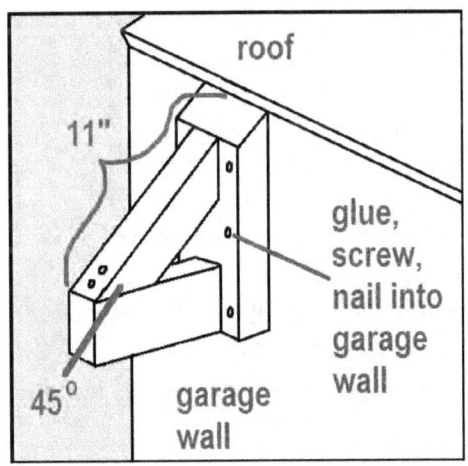

You may have to raise or lower the placement of the supports depending on how your roof sheathing meets the garage wall sheathing. You should caulk (or tar) any gap or seam, and also apply flashing.

You want the 1-ft wide plywood top to sit atop the supports and butt up to the roof or wall, but the lower edge should hang over the end of the support by 1" to 1/2". Or more than that depending on your width of facia. I used 1/2" plywood for my facia. If you used 2x4 for your facia, then you will need more of an overhang. And you may have to bevel the top and bottom edges of the 1-ft wide plywood top of the flareout to make ends flush and tight.

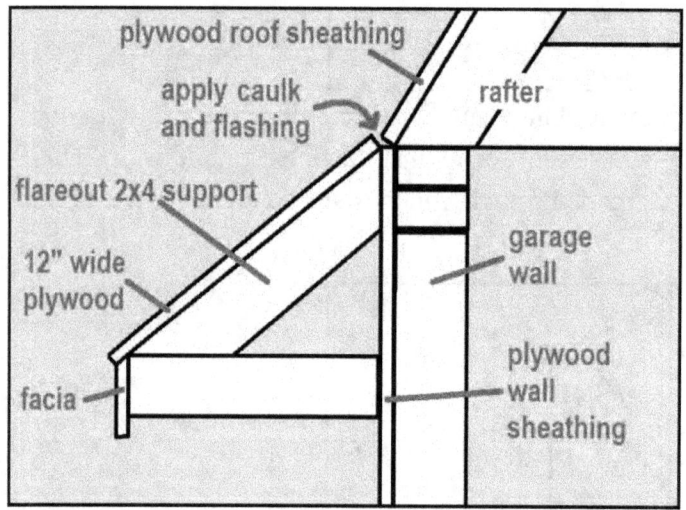

In any event, please use treated wood for your facia. I have seen too many houses/garages with rotted out facia. Attaching soffit up under the flare outs is optional and up to you.

Make sure the 1-ft wide flareout top extends over the last support enough so that it lines up with the roof sheathing.

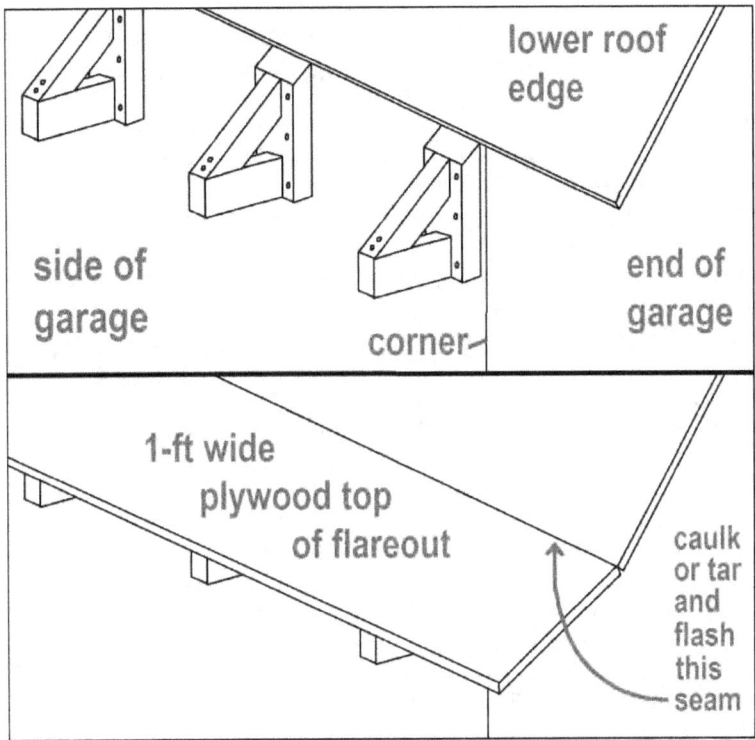

The number of supports will depend on how many studs are in your garage side walls. I put a support every place I had a wall stud. Strong flareouts might seem like overkill, but when you shingling the roof, having a flareout that can support your weight will be helpful.

Turkey tail

At the peak, on the back side, a decorative feature will be added, a turkey tail. A turkey tail is simply a triangular extension of the roof peak. This feature is sometimes also called crow's beak, bonnet, hay hood, widows peak, etc.

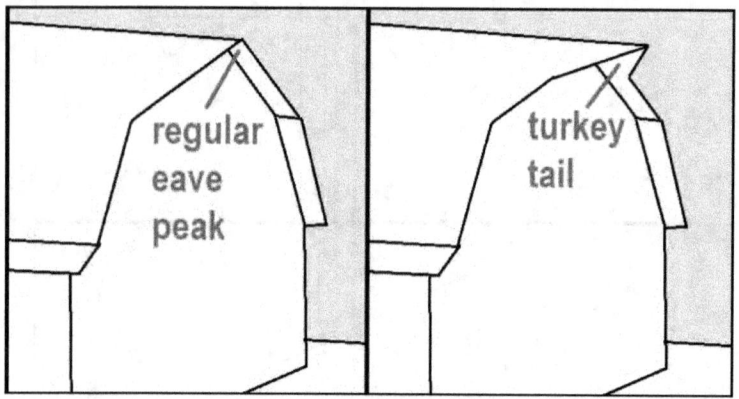

The eaves on the ends of the roof are just an extra foot of plywood hanging over the end walls. To make the turkey tail, we will add two 1-ft right-triangle pieces at the peak made from the same 1/2" plywood from which the roof is sheathed. The sides of the triangles are 1-ft and the hypotenuse is 1'5".

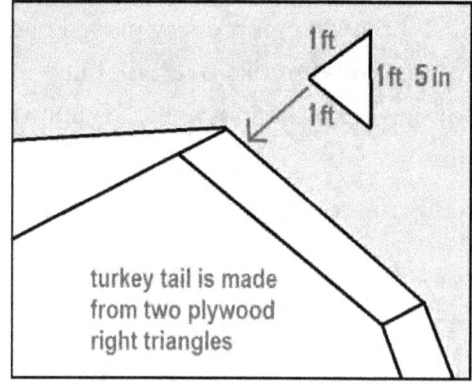

The corner with the 90° goes at the peak, so one of the 1-ft sides goes against the roof edge, while the other 1-ft side runs in-line with the roof peek (and will touch the second triangle. The longest side (hypotenuse) touches nothing. Use a plywood mending plate underneath the triangle to hold it in place.

Add the second triangle and its plywood mending plate.

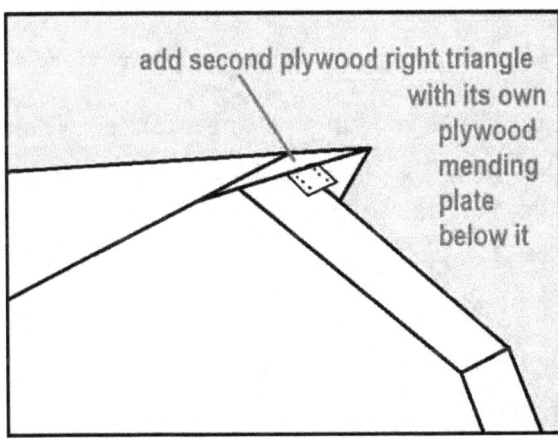

Use a bent metal mending plate below both of them (with screws) to connect the two triangles.

Facia block supports

The facia on the lower edge of the roof will be supported by the ends of the flareout supports, as explained in a previous section. However, the edges of the ends of the roof have nothing. The eaves on the ends of the roof are just an extra foot of plywood hanging over the end rafters. Facia and blocks need to be added.

Use lengths of treated 1x4 or 2x4 for the facia. Glue the top edge of the facia and clamp to the roof edge. Add screws through the top surface of the roof edge into the facia.

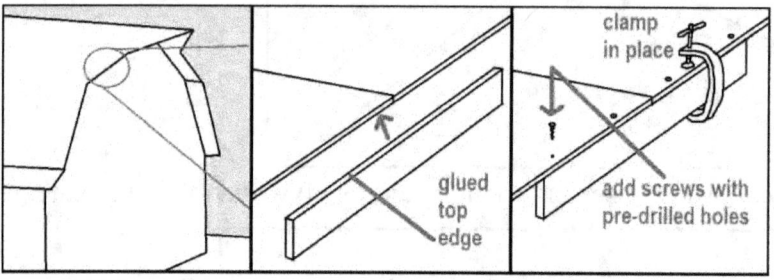

At the corners, where the roof angle changes, either use an angled-cut piece that spans the gap, or add a mending plate behind the seam of mitered facia pieces.

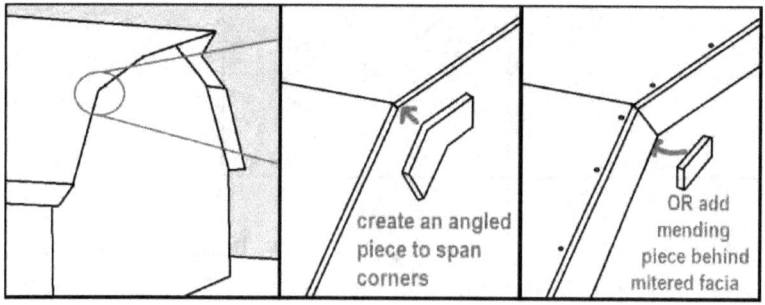

Once all the facia is in place on the ends of the roof, simple 2x4 blocks need to be added behind them for support. Add glue to the ends and top of the block. You can screw through the outside face of the facia into the outside ends of these blocks. Then toe nail the other end of the block into the outer roof section wall. You can also screw down through the roof into the blocks.

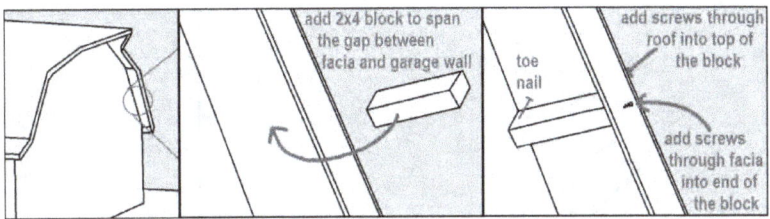

You can add soffit if you want. I didn't because I am cheap and lazy. I just let the wasps build their nests.

For decoration, in the lower corners (where the facia from the lower half of the gambrel roof meets the flareouts) I added a simple inward curve to the facia.

There are some building codes that require your facia to be "covered." That could mean they just need to be painted. However, you can also cover your facia with bent aluminum. Do not let treated wood touch the aluminum, or a chemical reaction can occur. Paint the treated wood or add a some tar paper between them. If you do not have access to a break to bend your aluminum, then you can bend it yourself over some angle iron.

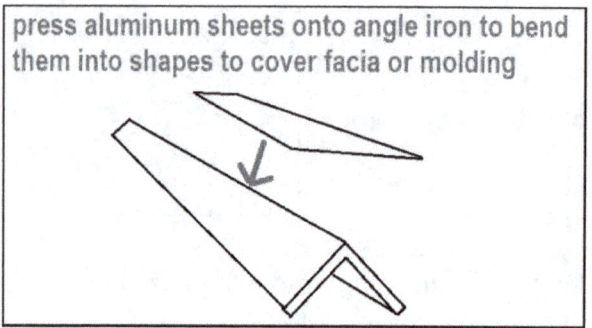

press aluminum sheets onto angle iron to bend them into shapes to cover facia or molding

Dormers

The last thing I did before shingling the roof was to build a small shed dormer in the middle of each side. Decide where you want your dormer. I chose the middle-most rafter.

shed dormer

Start by removing the 6-ft collar tie of the middle rafter.

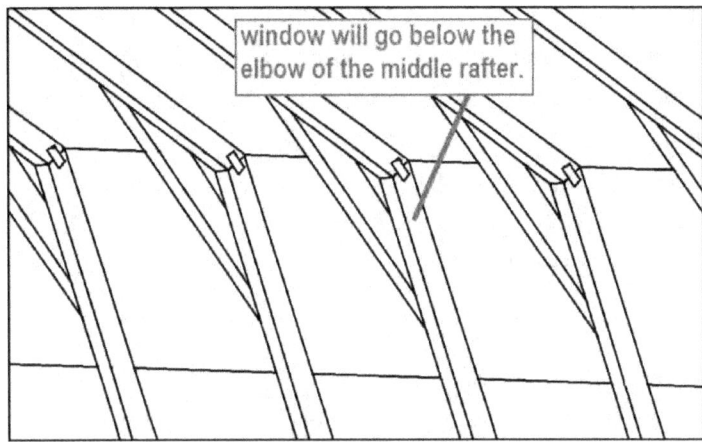

window will go below the elbow of the middle rafter.

If the collar ties on either outside rafter are on the inside of the rafter, then those ties need to be moved to the outside of that rafter.

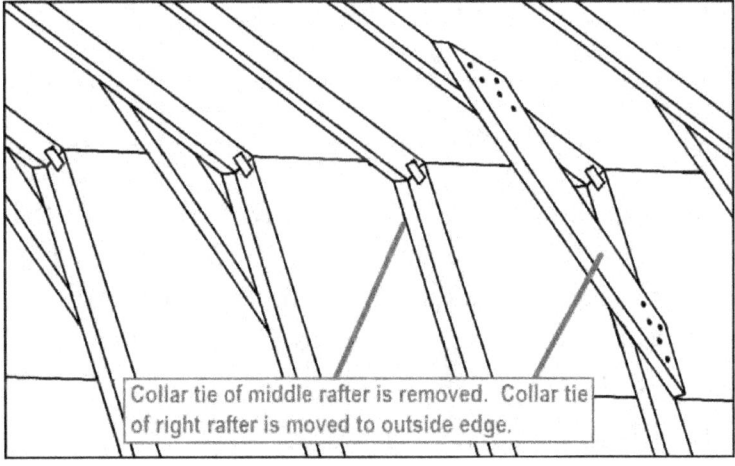

Collar tie of middle rafter is removed. Collar tie of right rafter is moved to outside edge.

Go down 3-ft from the center of the elbow (the seam where the upper roof meets the lower roof), or up 1-ft from the plywood seam of the roof sheathing, and draw a horizontal line on both sides of the middle rafter. Using a reciprocating saw, cut that line, and through the middle rafter, but stay 1-1/2" short of the two outside rafters. Then cut up vertically staying 1-1/2" away from the outside rafters. Later, this 1-1/2" edge will be filled with a trimmer stud.

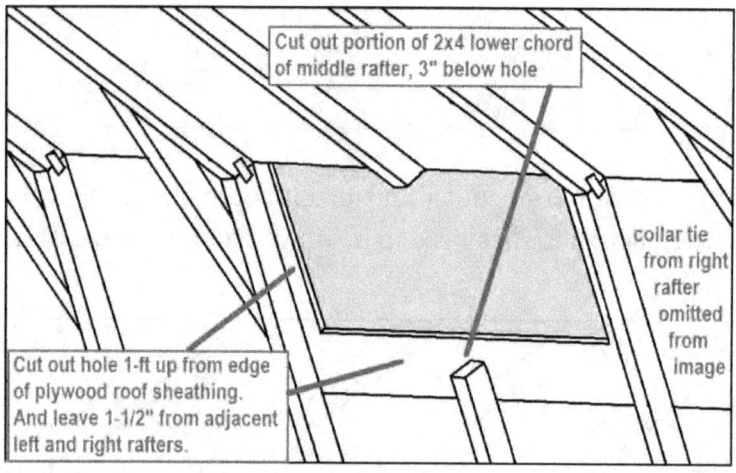

Cut out portion of 2x4 lower chord of middle rafter, 3" below hole

collar tie from right rafter omitted from image

Cut out hole 1-ft up from edge of plywood roof sheathing. And leave 1-1/2" from adjacent left and right rafters.

It may be difficult to remove this piece of the lower chord due to its connection to the upper chord and the roof sheathing nailed to it. Then cut the lower chord of the middle rafter again, 3" below the window hole. This 3" will be filled by a double sill plate in the next step.

Make sure you have 3 or 4 clamps with you for the rest of the steps. Various 2x4s components will be nailed together here and there.

Run two 2x4s from the inside edges of the outside rafters along the bottom of the window hole to create a sill plate. Affix cripple studs below this sill plate. Partial cripples will work in place of full cripples.

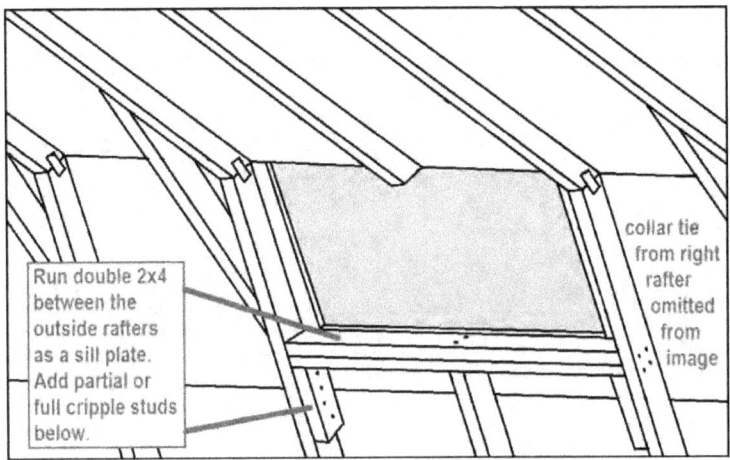

2x4 extensions will be added to the insides of the upper chords of the outer rafters. So two slots for them need to be cut out of the upper corners of the big hole.

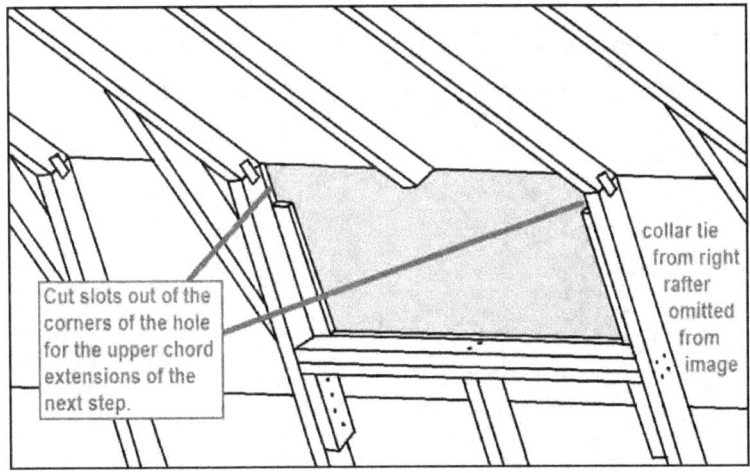

Four 2x4 extensions are affixed to the three upper chords. One to each outer rafter, and two to the middle rafter. These will support the top of the dormer.

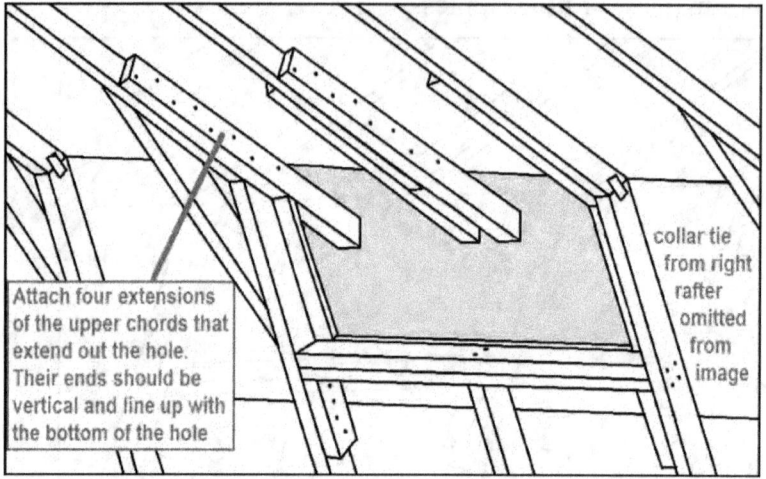

Attach four extensions of the upper chords that extend out the hole. Their ends should be vertical and line up with the bottom of the hole

collar tie from right rafter omitted from image

The outside ends of all four extensions need an angle cut, so their edge is vertical and plumb. And their cut ends line up with the bottom the hole. It is easier to make these cuts to the four 2x4 extensions ahead of time and then slide them out until they are lined up, then glue, screw, nail them to their upper chords. Again, having some clamps will make this task

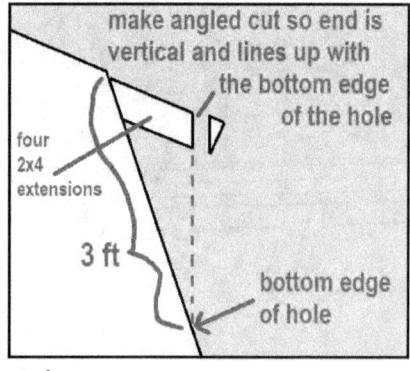

make angled cut so end is vertical and lines up with the bottom edge of the hole

four 2x4 extensions

3 ft

bottom edge of hole

easier.

Next, slide in two (one on each side) 2x4 trimmer studs next to the lower chords of the outer rafters. They sit atop the double 2x4 sill plate and, with angled-cut tops, nestle in under the two outer extensions.

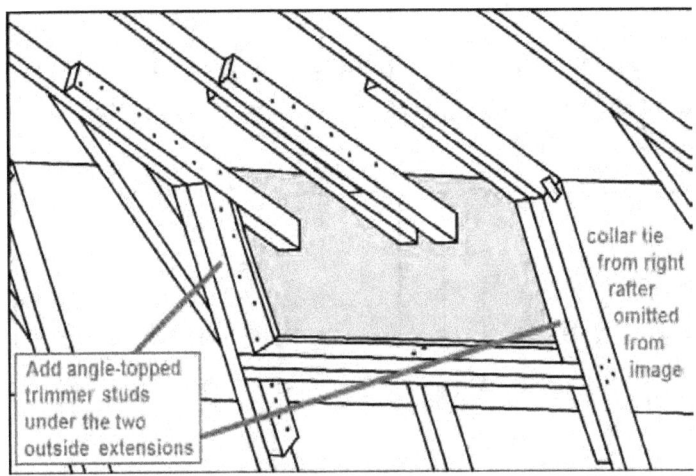

collar tie from right rafter omitted from image

Add angle-topped trimmer studs under the two outside extensions

A plywood face will go on the front of the dormer. It will nail into the ends of the four extensions. A bottom piece and two side pieces need to be added to for more support. The bottom piece is a 2x4 running the length of the window, and is added atop the double 2x4 sill plate. Its outside edge needs a long angle cut to make that surface vertical (plumb) and flush with the end edges of the four extensions.

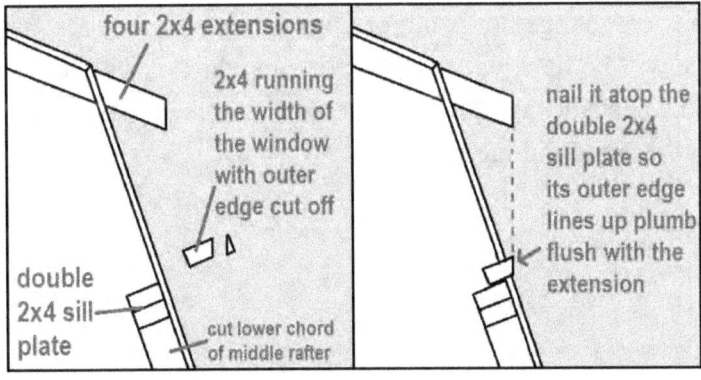

four 2x4 extensions

2x4 running the width of the window with outer edge cut off

double 2x4 sill plate

cut lower chord of middle rafter

nail it atop the double 2x4 sill plate so its outer edge lines up plumb flush with the extension

The two side pieces make up the dormer side. Along with the extensions they complete a triangle on each side. The front vertical piece is a 2x4 with angled-cut ends, with its bottom end resting on the roof surface. It can be attached via toe-nailed screws into the roof and extensions.

The longer piece is a 2x2 with angled-cut ends, with its long bottom side resting on the roof surface. Glue, screw, nail it into the roof surface.

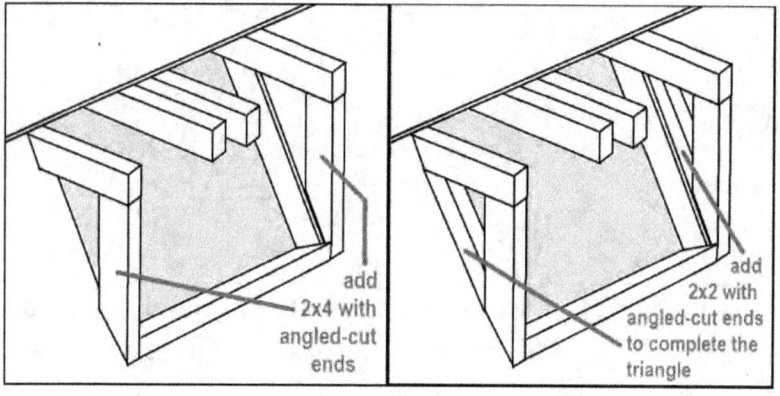

add
2x4 with
angled-cut
ends

add
2x2 with
angled-cut ends
to complete the
triangle

You could put a few 2x4 blocks between the ends of the extensions, but they are not necessary.

Time to put on the 1/2" plywood surfaces to complete the shed dormer. This comes in three parts: face, sides, and top. How they come together is important for weather/rain reasons. The face will be flush with the supports to which it will be attached. The sides will overlap the face, and then the top will overlap the sides and face.

Cut a rectangular piece of 1/2" plywood of the exact dimensions that match: 1. the top of the extensions, 2. the outer edge of side supports, 3. long bottom edge of the window opening. This piece will be the face of the dormer. The bottom edge of this plywood face must be beveled to fit the slope of the lower roof.

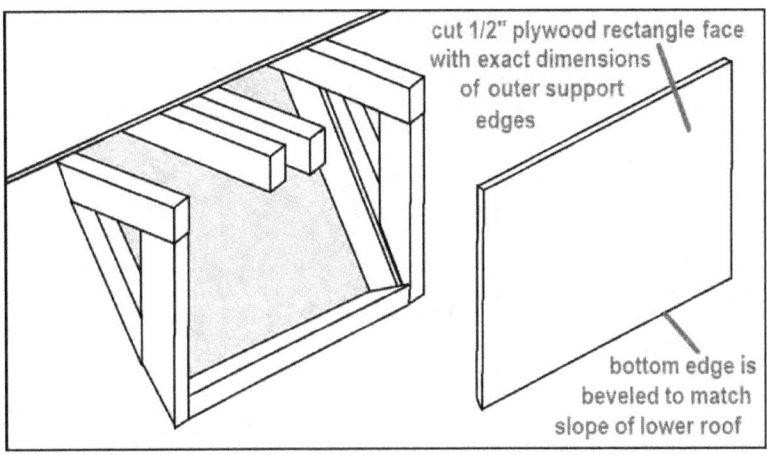

The illustration above shows a solid piece of plywood. You could cut for the hole for your window ahead of time if you wish. Glue, screw, nail the face to the supports.

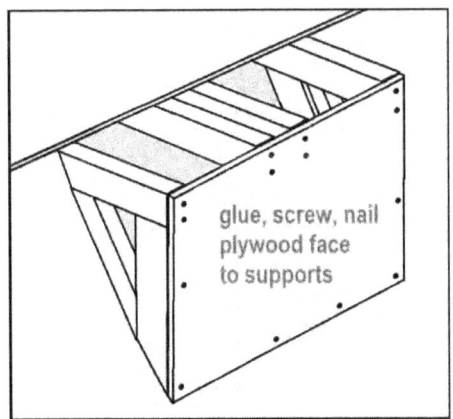

75

Next, cut two triangular plywood pieces that exactly match: 1. top of the extension, the lower roof, and the outside edge of the face.

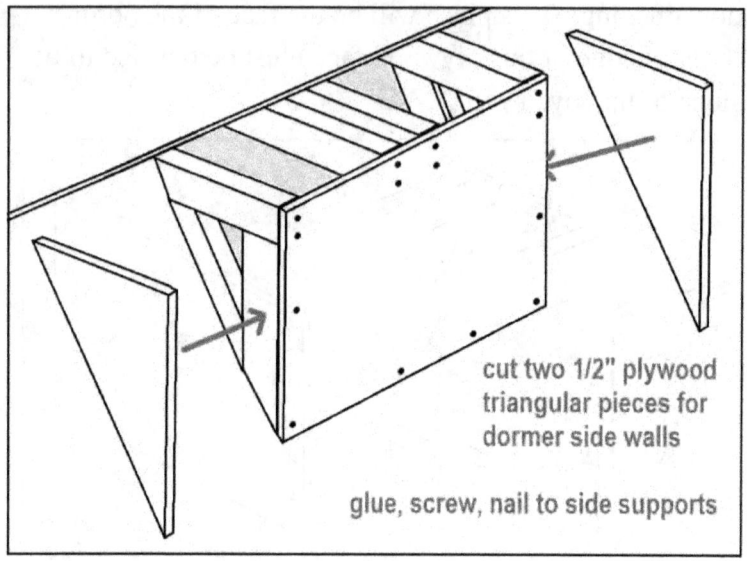

cut two 1/2" plywood
triangular pieces for
dormer side walls

glue, screw, nail to side supports

So the triangular piece will overlap the face edge.

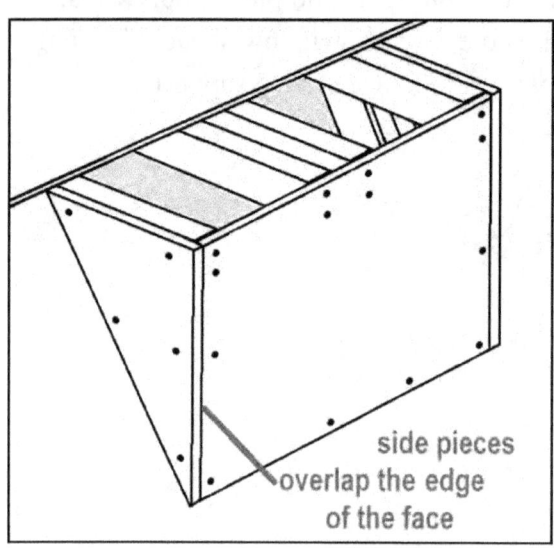

side pieces
overlap the edge
of the face

Last is the 1/2" plywood top. It's dimensions should exceed the top surface by just a little so that it overlaps the top edges of the face and sides.

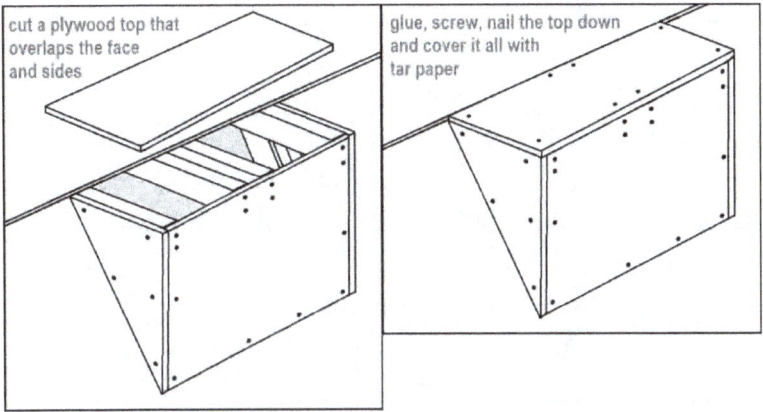

cut a plywood top that overlaps the face and sides

glue, screw, nail the top down and cover it all with tar paper

Caulk (or tar) the seams and add flashing. Then cover it all with tar paper.

For a window I found some cheap little spare 10"x12" window sills. I cut a hole and added some scrap 2x4 on the sides and built a little tilted ledge for a rough sill. Screwed on some hinges and a latch and called it good.

Once the flare outs, facia and blocks, turkey tail, and shed dormers are all done then the roof can be shingled.

Garage windows

I did not go out and buy $100 windows for the garage. They were going to go high (at least 5-ft off the floor) and under the eaves (flareouts) so I did not need them water tight. I found some lone 1'x2' sashes from an old window at a repurpose home-supply store that you find in alot of cities nowadays. Cut a horizontal rectangle for them, framed it like you normally would, and installed the sash with some hinges.

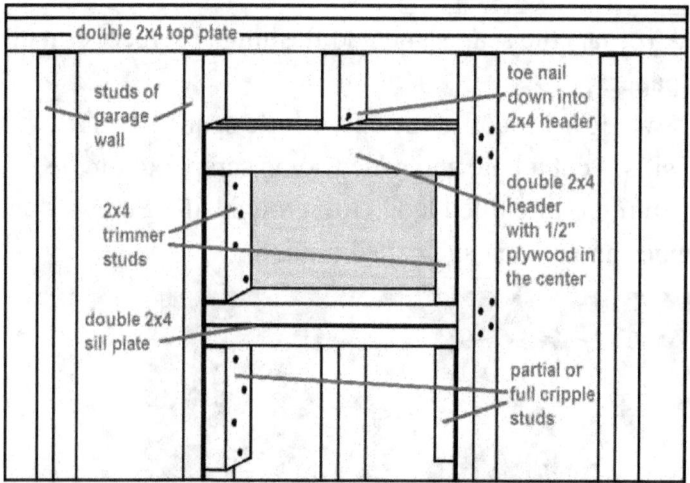

The cripple studs below the sill plate can just be partials or full. Full would run all the way to the bottom plate. Partials are just 1-2 ft long and affixed securely: glue, screw and nail.

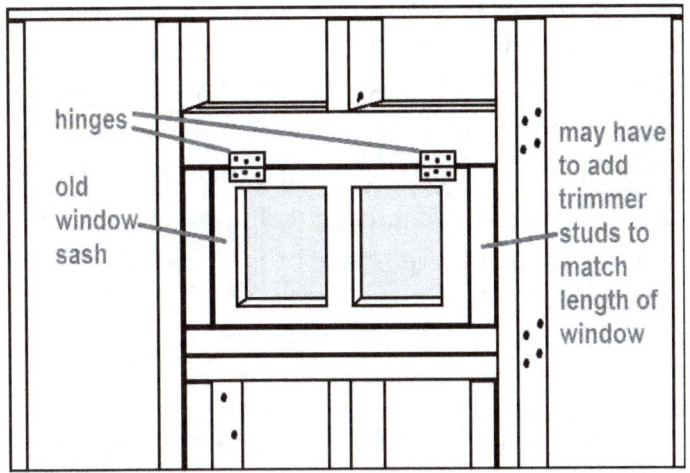

My old window does not seal perfectly, but I'm not worried about drafts or rain getting in.

Beam and pulley

The reason I wanted a gambrel-roofed garage (barn) is for the storage space of the upper level (roof section). I was not going to build stairs up to the upper level, so I knew I would need to set a pulley to lift heavy stuff up to the upper-level door. The pulley would be on a cantilevered 6x6 beam over the door.

This beam would be under the decorative turkey tail, so now the turkey tail has the purpose to protect that beam from the weather.

A 6x6 beam has the actual dimensions of 5-1/2" squared. On the inside of the back end wall, on the 2x6 kingpost of the end rafter, draw a line 5-1/2" down from the bottom edge of the top collar tie of that rafter. Cut away that portion of the 2x6 kingpost at that line and cut again just below the collar tie. Then cut a 5-1/2" square hole through the plywood sheathing where the portion of that kingpost had been. This is the hole that the 6x6 beam will slide through.

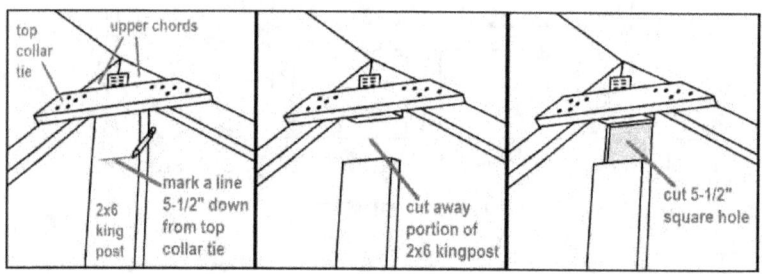

Despite the turkey tail above, the 6x6 beam will be exposed to the elements, so it needs to be treated wood. Slide the 6x6 beam through from the outside until it touches the 2x6 king post of the second rafter. It should hit right underneath the upper collar tie of that rafter. Glue, screw, nail a 2x6 cleat underneath the beam to hold it in place.

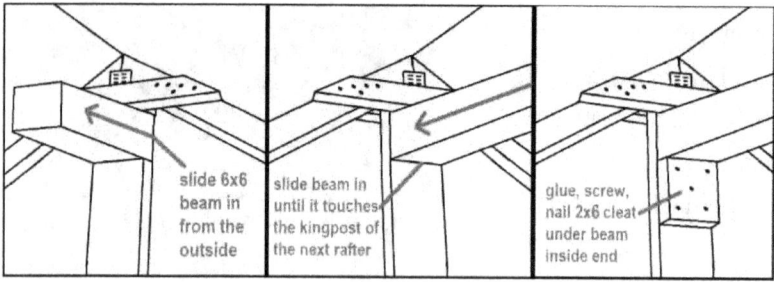

To secure the beam to that second rafter, screw big lag screws (with washers) through the other side of the rafter into the beam. Also add a 2x6 cleat to the 2x6 king post of the end rafter. If you make your cleats full-length (running all the way to the floor) all the better.

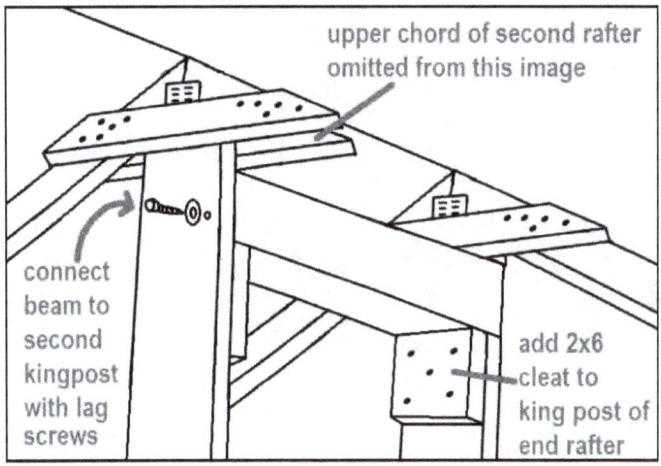

On the outside use long lag bolts to add a 6x6 45° brace under the beam. The outside end of my beam has an angle cut for decoration. Then add eye-hooks for your pulley.

I added a wooden rope cleat on the outside garage wall under the pulley to tie off ropes. The cleat is just a 12" 2x4 atop of a 4" 2x4. Be sure to add 2x4s to the inside of the wall so the two lag screws of the cleat have a good foundation.

In my build, I had removed the 2x6 king post of the second rafter to provide a doorway from one side of the upper level to the other side. So my 6x6 beam was affixed to the third rafter in. And the 2x6 king post I cut away from the second rafter became the cleat under the beam, and runs all the way to the floor.

Ladder

Having stairs up to the second floor would be nice, but they would just taken up too much room. Also, I don't go up there very often, so the convenience of stairs was not necessary. So I put in a simple ladder between some wall studs. For rungs I used 2x2s and I rounded their long edges some with a knife and sand paper. 16p nails were hammered through the 2x4s into the center of the 2x2s. The nails should go through pre-drilled holes so nothing will split.

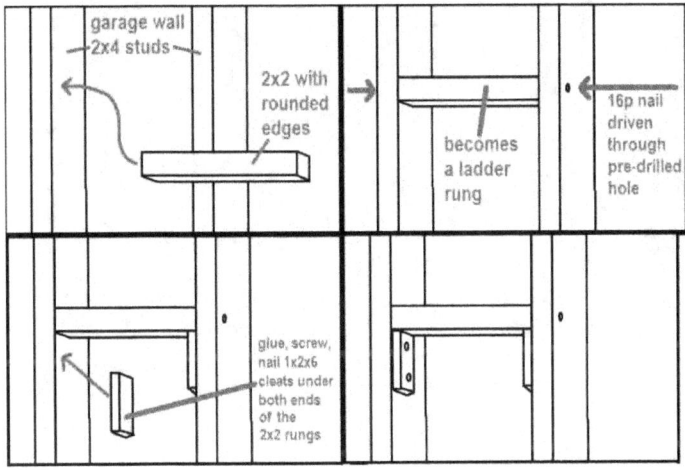

To make sure all the weight of someone climbing the ladder is not on just those two 16p nails, a 1x2x6 plywood cleat is glued, screwed, nailed under each rung.

I cut away a section of the plywood floor sheathing between two of the joists for a trap door. The ladder continues between two wall studs on the second floor so there is continual climbing through the trap door.

Siding

I sided my garage/barn with 4x8 sheets of beaded "car siding." It is thin plywood with grooves to look like bead board or shiplap. It is more like paneling than plywood. I nailed it on with ring-shank nails and painted it all red.

Hayloft door

On the upper level, there is a "hayloft door" just off the center of the end wall. The door was just plywood with 1x4s glued and screwed on as stiles and rails. Simple triangular strap hinges attach the door to the outer garage-end wall.

The plywood portion of the door fits in the hole of the door, but the stiles and rails do not, and extend an inch beyond the area of the plywood. So if the hole is 32" wide, then the plywood of the door is 31", and the stiles and rails are 34". Because of this, the stiles and rails part rest outside/atop the outer-most surface of the wall.

2x4 with beveled top added to keep rain from getting behind the door

So rain hitting the wall above the door would run down and get behind the door. So a 2x4 with a beveled top was affixed above the door.

The hinges screwed to the surface of the long side stile were not flush with the garage wall, so three 1x4 pieces were added just to the left of the door.

The door is held shut with a self-closing gate latch. The problem is, there is no way I can pull the handle to release the latch when the door is closed. So I simply drilled a descending hole through the wall and slid down a stiff wire. The outside end of the wire ties to the handle of the latch, and the inside end has a big loop for me to pull on.

The hole is descending so the weight of the wire will push the handle and close the latch.

Walk through

The second floor is separated in half by all the 2x6 king posts running down the center line. I made a big door way in the center by cutting two of those king posts free, then sliding them out to their neighbor king posts where they would be trimmer studs. A double 2x8 header sits above the trimmer studs.

Roof Vents

Like any roof, the very top of the roof section needs to be vented to let hot air out that builds up on summer days. I need to add some vents because I did not apply a ridge vent when I shingled the roof. The reason I left out the ridge vent is because I wanted to control the venting. I wanted the structure vented in the summer, but closed in the winter.

So I simply cut some rectangular holes near the top on both sides of the 2x6 king post on both end walls. They have metal screens that are stapled over them. In the winter I plug these screened holes with a piece of insulation to help keep the heat in.

<u>Hex sign</u>

Some people add quilts to their barns, mine has a hex sign.
Cut a 24" circle of painted aluminum and paint up a design
that is reflective of you. Mine was made by my mother. It has
Irish shamrocks, Dutch windmills, and a rose star.

Photo of my mother, who died the year before this book was finished.

IN CONCLUSION

Quite a few people told me building my garage in this way (backwards) was "crazy." But for me, I didn't really have a choice. I wanted to build it myself, and assembling the roof section on the ground was much safer than lifting those big rafters up onto the walls and nailing them in place. And then I would have had to lift full sheets of plywood up to sheath the roof. I wanted a safer build, and backwards was the best option.

Is this garage/barn strong and sturdy? Well, google search "2008 Cedar Rapids flood" and "2020 Iowa derecho." The structure survived direct hits by both.

If it looks daunting, just remember, the only things that are impossible, are things that have never happened before. If I did it, you can do it too. And if you are ever in Cedar Rapids, Iowa, stop by and I'll show the garage/barn to you.

About the author. Cal was born and raised in Iowa, graduated from Univ of Iowa, then had several jobs. In 2006 he bought a 100-year-old victorian fixer-upper.

When using a bottle jack on a cement block, be sure to put some plywood between the two, or you will end up getting stitches like me.

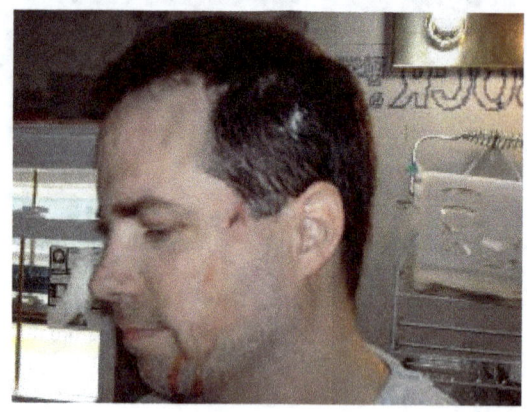